50 SCIENTIFIC DISCOVERIES THAT CHANGED THE WORLD

© 2014 Michael Wenkart

ISBN 9783735724991

„Herstellung und Verlag: BoD – Books on Demand, Norderstedt"

Bibliografische Information der Deutschen Nationalbibliothek: Die Deutsche Nationalbibliothek verzeichnet diese Publikation in der Deutschen Nationalbibliografie; detaillierte bibliografische Daten sind im Internet über **www.dnb.de** abrufbar.

Contents

1. The Big Bang Theory 1
2. Human Anatomy 2
3. Galileo's Law of Fall 5
4. Kepler's Laws of Planetary Motion 8
5. Newton's Laws of Motion 16
6. The Discovery of Oxygen 20
7. The Doctor Who Discovered Vaccines 32
8. Anaesthesia 36
9. The Discovery of the Atom 38
10. Levers and Buoyancy 41
11. Boyle's Law 45
12. The Discovery of the Cell 48
13. Discovery of Bacteria 52
14. The Ocean Controls the Weather 55
15. The Heliocentric Model 58
16. Satellites of Jupiter 61
17. Infrared and Ultraviolet 63
18. Relativity 66
19. The Doppler Effect 69
20. The Heart and the Circulatory System 71
21. Who Discovered Electricity? 76
22. Theory of Evolution 81
23. The Discovery of Radio Waves 87
24. Splitting the Atom 92
25. Discovery of DNA Structure and Function:
    Watson and Crick 99

26. Discovery of Antibiotics 110
27. The Founding of Genetics 117
28. The Microprocessor...127
29. Quantum Theory...135
30. The Invention of the Printing Press 139
31. A short history of the internal combustion engine..........143
32. The Invention of Paper 153
33. The Invention of The Internet......................................157
34. The Invention of The Computer 162
35. The Development of the Motor Car............................170
36. The Invention of The Radio 181
37. Invention of the Telephone 185
38. The Invention of Television 189
39. Invention of Paper Money...193
40. The Invention of the Aeroplane...................................197
41. The development of Cement...208
42. The Abacus 212
43. The History of the Nail 220
44. The Birth of Personal Computer 224
45. Assembly Line Production 228
46. The Cotton Gin 233
47. Hygiene and Sanitation..237
48. A Short History of Steel: Part II 245
49. The Steam Turbine...250
50. The Human Genome: A New Reality..........................258

50 Scientific Discoveries That Changed The World

Humanity has come a long way since our ancestors descended from the trees 15 million years agoand started the process that led to walking upright, developing tools, thinking strategically and, ultimately, the greatest advance of all – speech.

Progress isn't an even, ever-rising straight line though. It comes in fits and starts and is brought about by sudden great insights, ideas or observations that take evolution and development in important new directions. We will never know exactly how some of the great advances were made – speech, referred to above, obviously evolved in response to some great need of our very ancient forebears probably in relation to hunting; but we will never know for sure. Its true origin is lost in the mists of time. The wheel and fire were two other innovations that must have changed the lives of our ancestors immeasurably but, again, our knowledge of their exact provenance is limited.

More sophisticated adaptations like the development of tools, cultivation of crops and domestication of animals also had a profound effect on human evolution; but what we are describing in this book are fifty scientific discoveries or developments that we can attribute to specific persons or peoples that have transformed the way humans viewed the world and contributed to significant advances in the human condition. Whether they made life easier or longer, contributed to our expanding brains and knowledge or gave us understanding of the way the world (or even the universe) works – they have all benefitted the world and the people who live there in momentous ways.

We would not have vaccines, computers, motor cars, televisions, weather forecasting, industrial processes and production and a million other things without these inventions or discoveries. It is impossible to rank their importance – and we have not attempted to do so; but each one has contributed in their own way to who and where we are today as a species. You can make your own mind up as to which are the greatest and most important.

1

THE BIG BANG THEORY

The most popular theory of our universe's origin centers on a cosmic cataclysm unmatched in all of history—the big bang. This theory was born of the observation that other galaxies are moving away from our own at great speed, in all directions, as if they had all been propelled by an ancient explosive force.

Before the big bang, scientists believe, the entire vastness of the observable universe, including all of its matter and radiation, was compressed into a hot, dense mass just a few millimeters across. This nearly incomprehensible state is theorized to have existed for just a fraction of the first second of time.

Big bang proponents suggest that some 10 billion to 20 billion years ago, a massive blast allowed all the universe's known matter and energy—even space and time themselves—to spring from some ancient and unknown type of energy.

The Big Bang Theory

The theory maintains that, in the instant—a trillion-trillionth of a second—after the big bang, the universe expanded with incomprehensible speed from its pebble-size origin to astronomical scope. Expansion has apparently continued, but much more slowly, over the ensuing billions of years.

Scientists can't be sure exactly how the universe evolved after the big bang. Many believe that as time passed and matter cooled, more diverse kinds of atoms began to form, and they eventually condensed into the stars and galaxies of our present universe.

Origins of the Theory

A Belgian priest named Georges Lemaître first suggested the big bang theory in the 1920s when he theorized that the universe began from a single primordial atom. The idea subsequently received major boosts by Edwin Hubble's observations that galaxies are speeding away from us in all directions, and from the discovery of cosmic microwave radiation by Arno Penzias and Robert Wilson.

The glow of cosmic microwave background radiation, which is found throughout the universe, is thought to be a tangible remnant of leftover light from the big bang. The radiation is akin to that used to transmit TV signals via antennas. But it is the oldest radiation known and may hold many secrets about the universe's earliest moments.

The big bang theory leaves several major questions unanswered. One is the original cause of the big bang itself. Several answers have been proposed to address this fundamental question, but none has been proven—and even adequately testing them has proven to be a formidable challenge.

http://science.nationalgeographic.com/science/space/universe/origins-universe-article/

2

Human Anatomy

Andreas Vesalius

Andreas Vesalius was born in Brussels in 1515. His father, a doctor in the royal court, had collected an exceptional medical library. Young Vesalius poured over each volume and showed immense curiosity about the functioning of living things. He often caught and dissected small animals and insects.

At age 18 Vesalius traveled to Paris to study medicine. Physical dissection of animal or human bodies was not a common part of accepted medical study. If a dissection had to be performed, professors lectured while a barber did the actual cutting. Anatomy was taught from the drawings and translated texts of Galen, a Greek doctor whose texts were written in 50 B.C.

Vesalius was quickly recognized as brilliant but arrogant and argumentative. during the second dissection he attended,

Vesalius snatched the knife from the barber and demonstrated both his skill at dissection and his knowledge of anatomy, to the amazement of all in attendance.

In 1537 Vesalius graduated and moved to the University of Padua (Italy), where he began a long series of lectures, each centered on actual dissections and tissue experiments.

Like all medical students, Vesalius had been trained to believe in Galen's work. However, Vesalius had long been troubled because so many of his dissections revealed actual structures that differed from Galen's descriptions.

In a lecture that was held in January 1540, for the first time in public, Vesalius revealed his evidence to discredit Galen and to show that Galen's descriptions of curved human thighbones, heart chambers, segmented breast bones, etc., better matched the anatomy of apes than humans. In his lecture, Vesalius detailed more than 200 discrepancies between actual human anatomy and Galen's descriptions. Time after time, Vesalius showed that what every doctor and surgeon in Europe relied on fit better with apes, dogs, and sheep than the human body. Galen, and every medical text based on his work, were WRONG.

Vesalius secluded himself for three years preparing his detailed anatomy book. He used master artists to draw what he dissected, blood vessels, nerves, bones, organs, muscles, tendons, and brain.

Vesalius completed and published his magnificent anatomy book in 1543.

His published book became the standard anatomy text for over 300 years.

http://greatest-science-discoveries.blogspot.co.uk/2011/02/human-anatomy.html

3

GALILEO'S LAW OF FALL

Galileo Galilei

Falling Bodies, objects moving downward under the influence of gravity. The nature of this motion is the same for an object that falls straight down as it is for one that moves forward and down at the same time. Thus a bullet fired horizontally from a rifle falls at the same rate as one that is simply dropped. Knowledge of the motion of falling bodies is important in calculating the trajectory of bombs, bullets, artillery shells, and missiles.

In the fourth century B.C., Aristotle maintained that an object falls with a speed proportionate to its weightthat is, the heavier the object, the faster it falls. This idea was generally accepted until the 17th century, when Galileo showed that the rate of fall caused by gravity is the same for all objects. (There is a tradition that Galileo made this discovery by dropping iron balls of unequal weight from the Leaning Tower of Pisa. In describing his experiments, however, Galileo did not mention the Leaning Tower.) In an experiment made with balls rolling down a sloping board, Galileo determined the rate at which bodies accelerate (speed up) as they fall.

A seeming contradiction of the principle that all bodies fall at the same rate is the fact that a lump of lead will fall faster than such objects as feathers or leaves. However, these objects fall at a different rate because of air resistancein a vacuum, a lump of lead, a feather, and a leaf will fall at the same rate. This phenomenon was demonstrated by astronauts on the moon, which has no atmosphere. In one experiment, a hammer and a feather were dropped together from the same height; both fell at the same rate and struck the surface of the moon simultaneously.

In the mid-1980's, some physicists believed they had found evidence for a previously unknown force, much weaker than gravity, that would cause objects of different compositions to fall at very slightly different rates. Their report led to a number of experiments designed to detect such a force.

Law of Falling Bodies

The law of falling bodies states:

A falling body in a vacuum accelerates at the rate of 32 feet, per second (9.8 m/s) during each second that it falls. This acceleration is called the acceleration of gravity, which is expressed mathematically as g. (In air, the body accelerates

Galileo's Law of Fall

until it reaches its terminal velocitya constant velocity at which air resistance equals the force of gravity.)

The velocity (v) of a falling body that falls from rest is found by multiplying g by the time (t) during which a body falls: $v = gt$

For example, at the end of 5 seconds, a body will have a velocity of 32 X 5 = 160 feet per second.

The total distance (s) a body falls is equal to half of the acceleration of gravity multiplied by the square of the time: $S = gt2/2$

For example, at the end of the fifth second a body will have fallen

35 X 5 X 5/2 = 16X25 = 400 feet.

http://science.howstuffworks.com/falling-bodies-info.htm

4

Kepler's Laws of Planetary Motion

Johannes Kepler

Kepler

In the interplay between quantitative observation and theoretical construction that characterizes the development of modern science, we have seen that Brahe was the master of the first but was deficient in the second. The next great development in the history of astronomy was the theoretical intuition of Johannes Kepler (1571-1630), a German who went to Prague to become Brahe's assistant.

Brahe's Data and Kepler

Kepler and Brahe did not get along well. Brahe apparently mistrusted Kepler, fearing that his bright young assistant might eclipse him as the premiere astonomer of his day. He therefore let Kepler see only part of his voluminous data.

He set Kepler the task of understanding the orbit of the planet Mars, which was particularly troublesome. It is believed that part of the motivation for giving the Mars problem to Kepler was that it was difficult, and Brahe hoped it would occupy Kepler while Brahe worked on his theory of the Solar System. In a supreme irony, it was precisely the Martian data that allowed Kepler to formulate the correct laws of planetary motion, thus eventually achieving a place in the development of astronomy far surpassing that of Brahe.

Kepler and the Elliptical Orbits

Unlike Brahe, Kepler believed firmly in the Copernican system. In retrospect, the reason that the orbit of Mars was particularly difficult was that Copernicus had correctly placed the Sun at the center of the Solar System, but had erred in assuming the orbits of the planets to be circles. Thus, in the Copernican theory epicycles were still required to explain the details of planetary motion.

It fell to Kepler to provide the final piece of the puzzle: after a long struggle, in which he tried mightily to avoid his eventual conclusion, Kepler was forced finally to the realization that the orbits of the planets were not the circles demanded by Aristotle and assumed implicitly by Copernicus, but were instead the "flattened circles" that geometers call ellipses (See adjacent figure; the planetary orbits are only slightly elliptical and are not as flattened as in this example.)

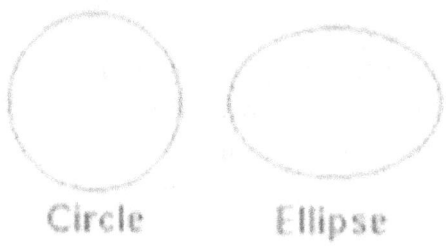

The irony noted above lies in the realization that the difficulties with the Martian orbit derive precisely from the fact that the orbit of Mars was the most elliptical of the planets for which Brahe had extensive data. Thus Brahe had unwittingly given Kepler the very part of his data that would allow Kepler to eventually formulate the correct theory of the Solar System and thereby to banish Brahe's own theory!

Some Properties of Ellipses

Since the orbits of the planets are ellipses, let us review a few basic properties of ellipses.

1. For an ellipse there are two points called foci (singular: focus) such that the sum of the distances to the foci from any point on the ellipse is a constant. In terms of the diagram shown to the left, with "x" marking the location of the foci, we have the equation

a + b = constant

that defines the ellipse in terms of the distances a and b.

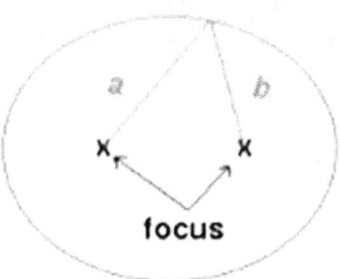

2. The amount of "flattening" of the ellipse is termed the eccentricity. Thus, in the following figure the ellipses become more eccentric from left to right. A circle may be viewed as a special case of an ellipse with zero eccentricity, while as the ellipse becomes more flattened the eccentricity approaches one. Thus, all ellipses have eccentricities lying between zero and one.

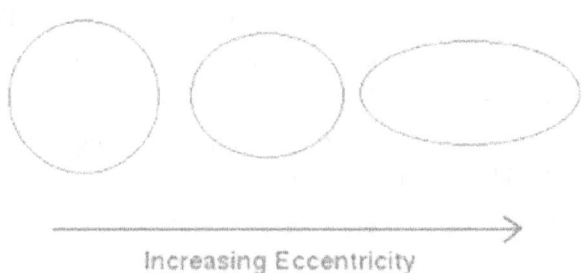

The orbits of the planets are ellipses but the eccentricities are so small for most of the planets that they look circular at first glance. For most of the planets one must measure the geometry carefully to determine that they are not circles, but ellipses of small eccentricity. Pluto and Mercury are exceptions: their orbits are sufficiently eccentric that they can be seen by inspection to not be circles.

3. The long axis of the ellipse is called the major axis, while the short axis is called the minor axis (adjacent figure). Half of the major axis is termed a semimajor axis. The length of a semimajor axis is often termed the size of the ellipse. It can be shown that the average separation of a planet from the Sun as it goes around its elliptical orbit is equal to the length of the semimajor axis. Thus, by the "radius" of a planet's orbit one usually means the length of the semimajor axis. For a more detailed investigation of the properties of ellipses, see this java applet.

The Laws of Planetary Motion

Kepler obtained Brahe's data after his death despite the attempts by Brahe's family to keep the data from him in the hope of monetary gain. There is some evidence that Kepler obtained the data by less than legal means; it is fortunate for the development of modern astronomy that he was successful. Utilizing the voluminous and precise data of Brahe, Kepler was eventually able to build on the realization that the orbits of the planets were ellipses to formulate his Three Laws of Planetary Motion.

Kepler's Law of Planetary Motion

Kepler's First Law:

I. The orbits of the planets are ellipses, with the Sun at one focus of the ellipse.

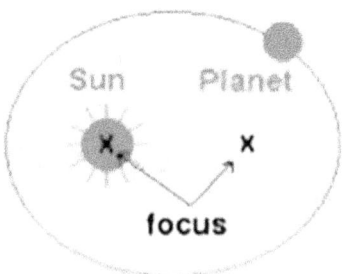

Kepler's First Law is illustrated in the image shown above. The Sun is not at the center of the ellipse, but is instead at one focus (generally there is nothing at the other focus of the ellipse). The planet then follows the ellipse in its orbit, which means that the Earth-Sun distance is constantly changing as the planet goes around its orbit. For purpose of illustration we have shown the orbit as rather eccentric; remember that the actual orbits are much less eccentric than this.

Kepler's Second Law:

II. The line joining the planet to the Sun sweeps out equal areas in equal times as the planet travels around the ellipse.

Kepler's second law is illustrated in the preceding figure. The line joining the Sun and planet sweeps out equal areas in equal times, so the planet moves faster when it is nearer the Sun. Thus, a planet executes elliptical motion with constantly changing angular speed as it moves about its orbit. The point of nearest approach of the planet to the Sun is termed perihelion; the point of greatest separation is termed aphelion. Hence, by Kepler's second law, the planet moves fastest when it is near perihelion and slowest when it is near aphelion.

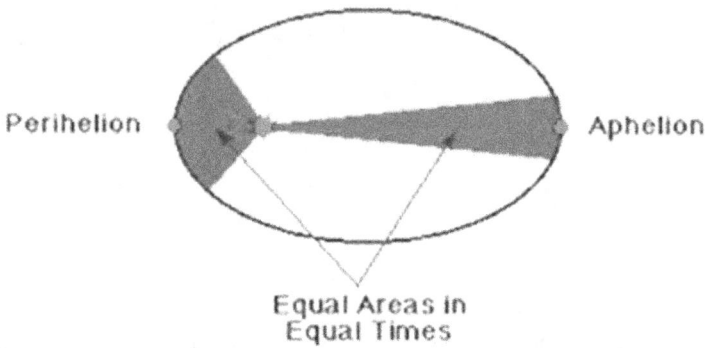

Equal Areas in Equal Times

Kepler's Third Law:

III. The ratio of the squares of the revolutionary periods for two planets is equal to the ratio of the cubes of their semimajor axes:

$$\frac{P_1^2}{P_2^2} = \frac{R_1^3}{R_2^3}$$

Kepler's Law of Planetary Motion

In this equation P represents the period of revolution for a planet and R represents the length of its semimajor axis. The subscripts "1" and "2" distinguish quantities for planet 1 and 2 respectively. The periods for the two planets are assumed to be in the same time units and the lengths of the semimajor axes for the two planets are assumed to be in the same distance units.

Kepler's Third Law implies that the period for a planet to orbit the Sun increases rapidly with the radius of its orbit. Thus, we find that Mercury, the innermost planet, takes only 88 days to orbit the Sun but the outermost planet (Pluto) requires 248 years to do the same.

http://csep10.phys.utk.edu/astr161/lect/history/kepler.html

5

Newton's Laws of Motion

Sir Isaac Newton

There was this fellow in England named Sir Isaac Newton. A little bit stuffy, bad hair, but quite an intelligent guy. He worked on developing calculus and physics at the same time. During his work, he came up with the three basic ideas that are applied to the physics of most motion (NOT modern physics). The ideas have been tested and verified so many times over the years, that scientists now call them Newton's Three Laws of Motion.

First Law

The first law says that an object at rest tends to stay at rest, and an object in motion tends to stay in motion, with the same direction and speed. Motion (or lack of motion) cannot change

Newton's Law of Motion

without an unbalanced force acting. If nothing is happening to you, and nothing does happen, you will never go anywhere. If you're going in a specific direction, unless something happens to you, you will always go in that direction. Forever.

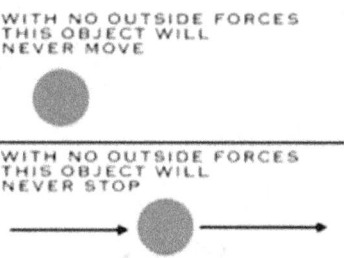

You can see good examples of this idea when you see video footage of astronauts. Have you ever noticed that their tools float? They can just place them in space and they stay in one place. There is no interfering force to cause this situation to change. The same is true when they throw objects for the camera. Those objects move in a straight line. If they threw something when doing a spacewalk, that object would continue moving in the same direction and with the same speed unless interfered with; for example, if a planet's gravity pulled on it (Note: This is a really really simple way of descibing a big idea. You will learn all the real details - and math - when you start taking more advanced classes in physics.).

Second Law

The second law says that the acceleration of an object produced by a net (total) applied force is directly related to the magnitude of the force, the same direction as the force, and inversely related to the mass of the object (inverse is a value that is one over another number... the inverse of 2 is 1/2). The second law shows that if you exert the same force on two objects of different mass, you will get different accelerations (changes in motion). The effect (acceleration) on the smaller mass will be greater (more noticeable). The effect of a 10 newton force on a baseball would be much greater than that same force acting on a truck. The difference in effect (acceleration) is entirely due to the difference in their masses.

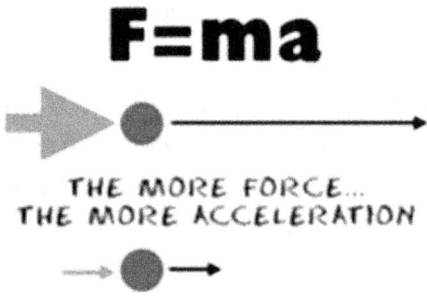

Third Law

The third law says that for every action (force) there is an equal and opposite reaction (force). Forces are found in pairs. Think about the time you sit in a chair. Your body exerts a force downward and that chair needs to exert an equal force upward or the chair will collapse. It's an issue of symmetry. Acting forces encounter other forces in the opposite direction. There's also the example of shooting a cannonball. When the cannonball is fired through the air (by the explosion), the cannon is pushed backward. The force pushing the ball out was equal to the force pushing the cannon back, but the effect on the cannon is less noticeable because it has a much larger mass. That example is similar to the kick when a gun fires a bullet forward.

http://csep10.phys.utk.edu/astr161/lect/history/kepler.html

6

THE DISCOVERY OF OXYGEN

Joseph Priestley

When Joseph Priestley discovered oxygen in 1774, he answered age-old questions of why and how things burn. An Englishman by birth, Priestley was deeply involved in politics and religion, as well as science. When his vocal support for the American and French revolutions made remaining in his homeland dangerous, Priestley left England in 1794 and continued his work in America until his death.

About Joseph Priestley

Some 2,500 years ago, the ancient Greeks identified air — along with earth, fire and water — as one of the four elemental components of creation. That notion may seem charmingly primitive now. But it made excellent sense at the time, and there was so little reason to dispute it that the idea persisted until the late 18th century. It might have endured even longer had it not been for a free-thinking English chemist and maverick theologian named Joseph Priestley.

Priestley (1733-1804) was hugely productive in research and widely notorious in philosophy. He invented carbonated water and the rubber eraser, identified a dozen key chemical compounds, and wrote an important early paper about electricity. His unorthodox religious writings and his support for the American and French revolutions so enraged his countrymen that he was forced to flee England in 1794. He settled in Pennsylvania, where he continued his research until his death.

The world recalls Priestley best as the man who discovered oxygen, the active ingredient in our planet's atmosphere. In the process, he helped dethrone an idea that dominated science for 23 uninterrupted centuries: Few concepts "have laid firmer hold upon the mind," he wrote, than that air "is a simple elementary substance, indestructible and unalterable."

In a series of experiments culminating in 1774, Priestley found that "air is not an elementary substance, but a composition," or mixture, of gases. Among them was the colorless and highly reactive gas he called "dephlogisticated air," to which the great French chemist Antoine Lavoisier would soon give the name "oxygen."

It is hard to overstate the importance of Priestley's revelation. Scientists now recognize 92 naturally occurring elements-including nitrogen and oxygen, the main components of air. They comprise 78 and 21 percent of the atmosphere, respectively.

Engraving of Joseph Priestley by Charles Turner, after a painting by Henry Fuseli.

Courtesy The Edgar Fahs Smith Memorial Collection, The University of Pennsylvania Libraries.

Understanding the Composition of Air

In the mid-18th century, the concept of an element was still evolving. Researchers had distinguished no more than two dozen or so elements, depending on who was doing the counting. It wasn't clear how air fit into that system. Nobody knew what it was, and researchers kept finding that it could be converted into such a variety of forms that they routinely spoke of different "airs."

The principal method for altering the nature of air, early chemists learned, was to heat or burn some compound in it. The second half of the 1700s witnessed an explosion of interest in such gases. The steam engine was in the process of transforming civilization, and scientists of all types were fascinated with combustion and the role of air in it.

British chemists were especially prolific. In 1754, Joseph Black identified what he called "fixed air" (now known to be carbon dioxide) because it could be returned, or fixed, into the sort of solids from which it was produced. In 1766, a wealthy eccentric named Henry Cavendish produced the highly flammable substance Lavoisier would name hydrogen, from the Greek words for "water maker."

Finally in 1772, Daniel Rutherford found that when he burned material in a bell jar, then absorbed all the "fixed" air by soaking it up with a substance called potash, a gas remained. Rutherford dubbed it "noxious air" because it asphyxiated mice placed in it. Today, we call it nitrogen.

But none of those revelations alone tells the whole story. The next major discovery would come from a man whose early life gave no indication that he would become one of the greatest experimental chemists in history.

Bubbling Beverages

In 1767, Priestley was offered a ministry in Leeds, Englane, located near a brewery. This abundant and convenient source of "fixed air" — what we now know as carbon dioxide —

from fermentation sparked his lifetime investigation into the chemistry of gases. He found a way to produce artificially what occurred naturally in beer and champagne: water containing the effervescence of carbon dioxide. The method earned the Royal Society's coveted Copley Prize and was the precursor of the modern soft-drink industry.

Oxygen and Other Discoveries in England

Joseph Priestley was born in Yorkshire, the eldest son of a maker of wool cloth. His mother died after bearing six children in six years. Young Joseph was sent to live with his aunt, Sarah Priestley Keighley, until the age of 19. She often entertained Presbyterian clergy at her home, and Joseph gradually came to prefer their doctrines to the grimmer Calvinism of his father. Before long, he was encouraged to study for the ministry. And study, as it turned out, was something Joseph Priestley did very well.

Aside from what he learned in the local schools, he taught himself Latin, Greek, French, Italian, German and a smattering of Middle Eastern languages, along with mathematics and philosophy. This preparation would have been ideal for study at Oxford or Cambridge, but as a Dissenter (someone who was not a member of the Church of England) Priestley was barred from England's great universities. So he enrolled at Daventry Academy, a celebrated school for Dissenters, and was exempted from a year of classes because of his achievements.

After graduation, he supported himself, as he would for the rest of his life, by teaching, tutoring and preaching. His first full-time teaching position was at the Dissenting Academy in Warrington. (Although obviously brilliant, original, outspoken and, by one report, of "a gay and airy disposition," Priestley had an unpleasant voice and a sort of stammer. That he made a living through lectures and sermons is further evidence of his extraordinary nature.)

In 1762, he was ordained and married Mary Wilkinson, the daughter of a prominent iron-works owner. She was, he noted, "of an excellent understanding, much improved by reading, of great fortitude and strength of mind, and of a temper in the highest degree affectionate and generous; feeling strongly for others and little for herself."

Priestley traveled regularly to London, and became acquainted with numerous men of science and independent thought, including an ingenious American named Benjamin Franklin, who became a lifelong friend. Franklin encouraged Priestley in his research, one result of which was The History and Present State of Electricity. For that work, and his growing reputation as an experimenter, Priestley was made a Fellow of the Royal Society in 1766.

The History book was too tough for a popular audience, and Priestley determined to write a more accessible one. But he could find no one to create the necessary illustrations. So, in typical fashion, he taught himself perspective drawing. Along the way, he made many mistakes, and discovered that India rubber would erase lead pencil lines — a fact he mentioned in the preface.

By the age of 34, Priestley was a well-established and respected member of Britain's scientific community. He was still paying a price for his religious nonconformity, however. When the explorer Captain James Cook was preparing for his second voyage, Priestley was offered the position of science adviser. But the offer was rescinded under pressure from Anglican authorities who protested his theology, which was evolving into a strongly Unitarian position that denied the doctrine of the trinity.

In retrospect, the Cook affair may have been all for the best. In 1773, the Earl of Shelburne asked Priestley to serve as a sort of intellectual companion, tutor for the earl's offspring, and librarian for his estate, Bowood House. The position provided access to social and political circles Priestley could never have gained on his own, while leaving ample free time for the research that would earn him a permanent place in scientific history.

He systematically analyzed the properties of different "airs" using the favored apparatus of the day: an inverted container on a raised platform that could capture the gases produced by various experiments below it. The container could also be placed in a pool of water or mercury, effectively sealing it, and a gas tested to see if it would sustain a flame or support life.

In the course of these experiments, Priestley made an enormously important observation. A flame went out when placed in a jar in which a mouse would die due to lack of air. Putting a green plant in the jar and exposing it to sunlight would "refresh" the air, permitting a flame to burn and a mouse to breathe. Perhaps, Priestley wrote, "the injury which is continually done by such a large number of animals is, in part at least, repaired by the vegetable creation." Thus he observed that plants release oxygen into the air — the process known to us as photosynthesis.

On August 1, 1774, he conducted his most famous experiment. Using a 12-inch-wide glass "burning lens," he focused sunlight on a lump of reddish mercuric oxide in an inverted glass container placed in a pool of mercury. The gas emitted, he found, was "five or six times as good as common air." In succeeding tests, it caused a flame to burn intensely and kept a mouse alive about four times as long as a similar quantity of air.

Priestley called his discovery "dephlogisticated air" on the theory that it supported combustion so well because it had no phlogiston in it, and hence could absorb the maximum amount during burning. (The year before, Swedish apothecary Carl Wilhelm Scheele isolated the same gas and observed a similar reaction. Scheele called his material "fire air." But his findings were not published until 1777.)

Whatever the gas was called, its effects were remarkable. "The feeling of it in my lungs," Priestley wrote, "was not sensibly different from that of common air, but I fancied that my breast felt peculiarly light and easy for some time afterwards. Who can tell but that in time, this pure air may become a fashionable article in luxury. Hitherto only two mice and myself have had the privilege of breathing it."

The Discovery of Oxygen

Phlogiston and Fire

In the mid-18th century, the most pressing issue in chemistry and physics was to determine what exactly happens when something burns. The prevailing theory was that flammable materials contained a substance called "phlogiston" (from the Greek word for burn) that was released during combustion.

The theory held that when a candle burned, for example, phlogiston was transferred from it to the surrounding air. When the air became saturated with phlogiston and could contain no more, the flame went out. Breathing, too, was a way to remove phlogiston from a body. A typical test for the presence of phlogiston was to place a mouse in a container and measure how long it lived. When the air in the container could accept no more phlogiston, the mouse would die.

The 19th century scientist Antoine Lavoisier disproved the existence of phlogiston and helped to form the basis of modern chemistry using Joseph Priestley's discovery of oxygen.

Pneumatic trough and other equipment used by Joseph Priestley in his experiments on oxygen and other gases. Reproduced from Joseph Priestley's book Experiments and Observations on Different Kinds of Air, 1774-1786.

Religion and Emigration to America

As luck would have it, Lord Shelburne was setting off on a trip to the European continent and took Priestley along. In France, Priestley met Lavoisier and described his discovery. It turned out to be the clue Lavoisier needed to develop his theory of chemical reactions — the "revolution" in chemistry that would finally dispel the phlogiston theory. Burning substances, Lavoisier argued, did not give off phlogiston; they took on Priestley's gas, which Lavoisier called "oxygen" from the Greek word for acid-maker.

By then, however, Priestley had returned to England, where he escalated his support for the American Revolution and for highly unorthodox religious views. Those positions were a source of embarrassment for Lord Shelburne. Priestley left his service in 1780, moving to Birmingham and taking a position as head of a liberal congregation called New Meeting.

His new location brought him into contact with numerous luminaries including Erasmus Darwin, grandfather of Charles, the great architect of evolutionary theory. James Watt and Matthew Boulton — who were about to transform society with their steam engine — were there, as was Josiah Wedgwood, the famous potter, who supported Priestley's chemical experiments. Birmingham also boasted a distinguished scientific discussion group, the Lunar Society, which met on nights of a full moon so that the members could see their way home.

Priestley's encouragement of the French Revolution, together with his increasingly controversial theology and attacks on the doctrine of the trinity, eventually became too notorious for safety. In 1791, an alcohol-fueled mob of royalists burned the New Meeting house, and then Priestley's home. The scientist and his family barely escaped. They fled to London, but eventually it proved no safer. Priestley's sons could not find work and emigrated to Pennsylvania, where they hoped to found a center for free-thinking Englishmen.

Finally Joseph and Mary followed them, setting sail for America on April 8, 1794. Priestley turned down the offer of a teaching position at the University of Pennsylvania in Philadelphia, and instead built a house in the remote hamlet of Northumberland to be near his sons. The area was decidedly rustic.

There Priestley continued his research, isolating carbon monoxide (which he called "heavy inflammable air") and founding the Unitarian Church in the United States. For the most part, he led a quiet and reflective life — especially after his friend Thomas Jefferson was elected president in 1800.

During his final trip to Philadelphia, he told the Philosophical Society that "having been obliged to leave a country which has been long distinguished by discoveries in science, I think myself happy by my reception in another which is following its example, and which already affords a prospect of its arriving at equal eminence." His words proved prophetic. A colloquium held on the centennial of Priestley's discovery of oxygen led to the founding of the American Chemical Society —today the world's largest scientific society — in 1876.

On February 3, 1804, Priestley began an experiment, but found himself too weak to continue. He went to his bed in his library, never again to emerge. On February 6, he summoned one of his sons and an assistant. He dictated some changes in a manuscript. When he was satisfied with the revisions, he said "That is right. I have now done." Minutes later he died painlessly, ending what Jefferson called "one of the few lives precious to mankind."

http://www.acs.org/content/acs/en/education/whatischemistry/landmarks/josephpriestleyoxygen.html

7

THE DOCTOR WHO DISCOVERED VACCINES

Edward Jenner

With all the media attention on new viruses and possible flu pandemics, it's easy to forget that some of the most devastating bugs ever to plague humankind have been wiped out.

Consider smallpox. The last lab samples sit under lock and key. But two centuries ago, just before an English country doctor named Edward Jenner stepped forward to attack it, smallpox killed people by the tens of thousands.

Dr. Jenner was born in a small town in Gloucestershire in 1749. While studying medicine in London, he was offered the opportunity to be the naturalist on one of Captain Cook's expeditions to the South Seas. He declined the offer because he preferred to practice medicine in the rural district where he had been raised.

And so he did. He loved the Gloucestershire landscape, and he studied its rocks, observed the migrations of its birds, and even built the first aerial balloon seen in those parts.

One thing consistent from Gloucestershire to the South Seas in Dr. Jenner's time was smallpox. It occasionally broke out in intense and lethal epidemics.

Other times it struck people at random. It claimed lives the world over. As with chickenpox today, people who had survived one case of smallpox were immune to it. Those who survived a bout of smallpox, however, were often hideously disfigured for life by pocklike scars. A primitive form of smallpox vaccination called "variolation" had gradually become popular in England, where "matter" from a skin pox of a smallpox-infected individual was inoculated into an uninfected person's skin. Most of the time, the inoculated person got a very mild case of smallpox and seemed to be protected from getting severe smallpox. But about 2 percent of people variolated got severe smallpox as a result of the procedure and died. Variolation was also problematic because it spread other diseases.

While Dr. Jenner treated country folk and played violin in the local music club, he thought about finding a cure for smallpox--and wondered if the answer was hidden in a bit of Gloucestershire lore.

The Doctor Who Discovered Vaccines

The popular belief held that a mild disease called cowpox kept away the dangerous smallpox. Cowpox was a disease of cows, but milkmaids occasionally caught it, and those who did almost never got smallpox.

Dr. Jenner first investigated this in 1775 and within five years he had satisfied himself that cowpox was usually caused by the same (or a similar) type of virus that caused human smallpox. Human cases of cowpox were scarce in Gloucestershire at the time, however, so Dr. Jenner did not get to test his idea until more than 20 years later.

In 1796, in one of the most famous scenes in medical history, Dr. Jenner took matter from the cowpox lesions on a milkmaid's finger and injected it into the skin of 8-year-old James Phipps. James developed symptoms of cowpox: a mild fever and a small skin lesion.

Six weeks later, Dr. Jenner inoculated James with smallpox. But James did not get the disease. The process of inoculating the boy with cowpox matter--a process Jenner called "vaccination" from the Latin word for cow (vacca)--had worked.

Dr. Jenner repeated his experiment in 1798, and he published his success, announcing it to the world.

After overcoming the initial skepticism of the medical establishment, the Royal Jennerian Society was established in 1803 to give proper vaccinations.

For the first time, vaccinations were done on a sweeping scale.

The Doctor Who Discovered Vaccines

The triumph paved the way for immunizations, such as the chickenpox vaccine.

Dr. Jenner died in 1823, having done more than any other person to wipe out one of history's most dreadful diseases.

The World Health Organization officially declared smallpox eradicated from the world in 1977, when the last case occurred in Africa. Its eradication was a result of mass vaccinations that began with Dr. Jenner's cowpox vaccine. The vaccine has remained basically unchanged since then.

http://www.urmc.rochester.edu/encyclopedia/content.aspx?ContentTypeID=1&ContentID=1061

8

Anaesthesia

We perform some pretty amazing and intricate surgeries today. heart bypasses, transplants and the list goes on. But we can't do any of it withoutanaesthesia.

There were crude forms of anaesthesia as early as 70 A.D. An early Roman physician named PedaniusDioscorides used opium and mandrake. In China there was and still is acupuncture. And around the world alcohol was used but vomiting was an unworkable side effect.

Serious early work with anaesthesia began with nitrous oxide and ether. An English chemist, discovered that nitrous oxide relieved his headache and dental pain, but his report went unnoticed.

Then an American dentist extracted his own teeth under nitrous oxide. But Horace Wells failed in a public demonstration because the anaesthesia bag was withdrawn early and the patient felt pain. Thus it did not catch on here or in Europe.

Others tried ether. In 1842, Crawford Long was the first to remove a tumor from a patient under ether but he didn't publish the event. Four years later Henry Bigelow operated with ether in a public demonstration so by 1847, ether and chloroform were firmly established as general anaesthetics.

Later, major advances include local anaesthesia using cocaine which led to infiltration anaesthesia such as nerve blocks, spinal and epidural anaesthesia. Next came control of the airway using tubes placed in the trachea to help breathing.

By the 1920's patients were given anaesthesia intravenously which helped them fall asleep quickly and pleasantly.

In the early 20th century ether and chloroform were replaced by halogenated hydrocarbons such as halothane which is still used.

Today we have this "balanced anaesthesia" approach: a "cocktail" of drugs to induce loss of consciousness and eliminate pain.

9

THE DISCOVERY OF THE ATOM

John Dalton was the first person to suggest the idea of the atom as being the smallest particle or building block that created all other materials that is around us. Since the introduction of this theory, numerous scientists of that time also came up, independently, with the idea of existing small particles. Later on, with the development of new scientific techniques, it was proven by the scientific community in the 19th century that John Dalton was right after all.

John Dalton

It took a long time to prove that the theory was right

Dalton had come up with this idea in 1803 and it was a lot earlier than when it was finally proven, scientifically, which was towards the end of the 19th century. Dalton only developed the idea of the atom, but he didn't know exactly what role it played in nature or in any other objects. The components that made up the atom were also not questioned as the general idea was that the atom is the smallest part of any matter.

Positive and negatively charged electrons

John Dalton's atomic symbols, representing all known elements that have been discovered by the scientific community at the time.

Many scientists, whom, due to geographical distance, couldn't have been in communication with each other, also developed theories about the existence of smaller particles, with positive and negatively charged electrons that keep the matter together. These positive and negatively charged electrons, which are now known as protons and electrons, were only acknowledged after the discovery of the atom.

But the history of the atom goes back even further

Latin physicists, Democritus and Leucippus have individually developed the idea of an atom and they even named it "atom", a Latin word meaning indivisible. So, the theory about the atom goes as far back as the 4th century B.C. Surprisingly, it took about 22 centuries later for somebody to figure out the existence of the small particle and another century more for somebody to discover how the atom works and how to experiment with it.

http://discovery.yukozimo.com/who-discovered-the-atom/

10

LEVERS AND BUOYANCY

Archimedes

The concepts of buoyancy (water pushes up on an object with a force equal to the weight of water that the object displaces) and of levers (a force pushing down on one side of a lever creates a lifting force on the other side that is proportional to the lengths of the two sides of the lever) lie at the foundation of all quantitative science and engineering.

They represent humanity's earliest breakthroughs in understanding the relationships in the physical world around us and in devising mathematical ways to describe the physical phenomena of the world. Countless engineering and scientific advances have depended on those two discoveries.

In 260 B.C. 26-year-old Archimedes studied the two known sciences, astronomy and geometry, in Syracuse, Sicily. One day Archimedes was distracted by four boys playing on the beach

Levels and Buoyancy

with a driftwood plank. They balanced the board over a waist-high rock. One boy straddled one end while his three friends jumped hard onto the other. The lone boy was tossed into the air.

The boys slid the board off-center along their balancing rock so that only one-quarter of it remained on the short side. Three of the boys climbed onto the short, top end. The fourth boy bounded onto the rising long end, crashing it back down to the sand and catapulting his three friends into the air.

Archimedes was fascinated. And he determined to understand the principles that so easily allowed a small weight (one boy) to lift a large weight (three boys).

Archimedes used a strip of wood and small wooden blocks to model the boys and their driftwood. He made a triangular block to model their rock. By measuring as he balanced different combinations of weights on each end of the lever (lever came from the Latin word meaning "to lift"), Archimedes realized that levers were an example of one of Euclid's proportions at work. The force (weight) pushing down on each side of the lever had to be proportional to the lengths of board on each side of the balance point. He had discovered the mathematical concept of levers, the most common and basic lifting system ever devised.

Fifteen years later, in 245 B.C., Archimedes was ordered by King Hieron to find out whether a goldsmith had cheated the king. Hieron had given the smith a weight of gold and asked him to fashion a solid-gold crown. Even though the crown weighed exactly the same as the original gold, the king suspected that the goldsmith had wrapped a thin layer of gold around some other, cheaper metal inside. Archimedes was ordered to discover whether the crown was solid gold without damaging the crown itself.

It seemed like an impossible task. In a public bathhouse Archimedes noticed his arm floating on the water's surface. A vague idea began to form in his mind. He pulled his arm completely under the surface. Then he relaxed and it floated back up.

He stood up in the tub. The water level dropped around the tub's sides. He sat back down. The water level rose.

He lay down. The water rose higher, and he realized that he felt lighter. He stood up. The water level fell and he felt heavier. Water had to be pushing up on his submerged body to make it feel lighter.

He carried a stone and a block of wood of about the same size into the tub and submerged them both. The stone sank, but felt lighter. He had to push the wood down to submerge it. That meant that water pushed up with a force related to the amount of water displaced by the object (the object's size) rather than to the object's weight. How heavy the object felt in the water had to relate to the object's density (how much each unit volume of it weighed).

That showed Archimedes how to answer the king's question. He returned to the king. The key was density. If the crown was made of some other metal than gold, it could weigh the same but would have a different density and thus occupy a different volume.

The crown and an equal weight of gold were dunked into a bowl of water. The crown displaced more water and was thus shown to be a fake.

More important, Archimedes discovered the principle of buoyancy: Water pushes up on objects with a force equal to the amount of water the objects displace.

When Archimedes discovered the concept of buoyancy, he leapt form the bath and shouted the word he made famous: "Eureka." which means "I found it." That word became the motto of the state of California after the first gold rush miners shouted that they had found gold.

http://www.bigsiteofamazingfacts.com/who-discovered-the-fundamental-principles-of-levers-and-buoyancy-in-physics-and-engineering/

11

BOYLE'S LAW

Robert Boyle

Boyle's Law is a basic law in chemistry describing the behavior of a gas held at a constant temperature. The law, discovered by Robert Boyle in 1662, states that at a fixed temperature, the volume of gas is inversely proportional to the pressure exerted by the gas. In other words, when a gas is pumped into an enclosed space, it will shrink to fit into that space, but the pressure that the gas puts on the container will increase. Boyle's Law can be written out mathematically:

P x V = constant

In this equation, P = pressure and V = volume.

Boyle's Experiment

To prove the law, Boyle pumped oxygen (a gas) into a J-shaped tube of glass that was sealed at one end. Using a burner to keep the oxygen at a constant temperature, he then poured different amounts of mercury into the tube, which varied the pressure on the oxygen. He found that the more pressure he applied, the smaller the volume of the oxygen, and this reduction happened at a constant rate.

Boyle's Law specifically relates to an ideal gas — that is, a theoretical gas that is made up of random particles that do not interact. Although no real gasses are ideal gasses, most do display these ideal characteristics under normal conditions.

Real-World Examples

One example of Boyle's Law in action can be seen in a balloon. Air is blown into the balloon; the pressure of that air — a gas — pushes on the rubber, making the balloon expand. If one end of the balloon is squeezed, making the volume smaller, the pressure inside increases, making the un-squeezed part of the balloon expand out. There is a limit to how much the gas can be compressed, however, because eventually the pressure becomes so great that it causes the balloon (or any container) to break.

A different example is a syringe for taking blood. An empty syringe has a fixed amount of gas (air) in it; if the plunger is drawn back without the needle end being inserted into anything, the volume of the tube will increase and the pressure will drop, causing more air to move into the tube to equalize the pressure. If the syringe is inserted into a vein and the plunger drawn back, blood will flow into the tube since the pressure in the vein is higher than the pressure in the syringe.

Another way of describing Boyle's law is that when pushed, a gas tends to push back. Without the massive amount of gravity holding them together, the solar system's gas planets would rapidly diffuse in all directions, quickly depressurizing. In this case, the pressure of gravity regulates the volume of the gases around these planets.

http://www.wisegeek.org/what-is-boyles-law.htm

12

THE DISCOVERY OF THE CELL

Robert Hooke

In 1661 King Charles II of England commissioned Sir Christopher Wren to create a series of microscopical studies. Wren obliged, but after a few presentations found he didn't have the time and gave up the project to an upcoming scientist with something of a knack for drawing and mechanics. The rest is history.

Robert C. Hooke (1635-1703) was 26 years old when he took the assignment from Wren and joined the Royal Society For Scientists. A self-educated child prodigy, he showed technical aptitude by recreating the entire inner workings of a clock out of wood, then assembling it to run. Hooke also taught himself technical drawing, a skill he used to capture observations through his microscope.[1]

The Discovery of The Cell

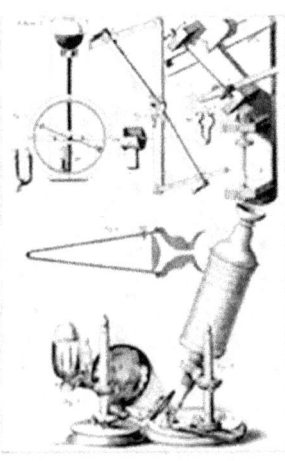

Hooke's Drawing of his Microscope from Micrographia.

Hooke applied his technical abilities to invent ways of controlling the height and angle of microscopes, as well as mechanisms of illumination. Variations in light allowed Hooke to see new detail, and he used multiple sources of illumination before producing any single drawing. Hooke's technical efforts created magnifications of 50x, enabling insight to a world not yet known in the 1600s.1

King Charles only requested insect studies, but Hooke went beyond his commission and looked at everything from fabric, leaves, mica, glass, flint, and even frozen urine. Hooke did things like let a louse suck from his hand to observe how his blood traveled through its innards. He also stung himself with nettles to see where and how the poison was pumped into his hands.1

The Discovery of The Cell

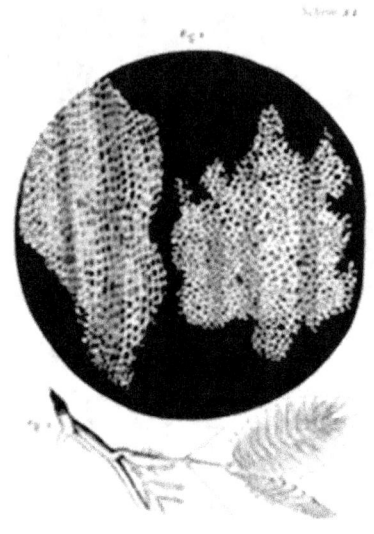

Hooke viewed a thin cutting of cork and discovered empty spaces contained by walls which he termed cells.

When Hooke viewed a thin cutting of cork he discovered empty spaces contained by walls, and termed them pores, or cells. The term cells stuck and Hooke gained credit for discovering the building blocks of all life. Hooke calculated the number of cells in a cubic inch to be 1,259,712,000, and while he couldn't grasp the full effect of his discovery, he did at least appreciate the sheer number of these cells.2

Hooke recorded all his drawings and observations into Micrographia: or Some Physiological Descriptions of Miniature Bodies Made by Magnifying Glasses. After acknowledging the King and the Royal Society, the book covered a wide range of topics from the construction of microscopes themselves, to the spectrum of color, the molecular causes of fire, the crystal structure of objects, and the anatomy of insects. Published in 1665, the book became an instant best seller. Hooke had ignited the spark of cell theory and set a trend of scientists making discoveries by looking through microscopes on government payroll.

http://www.science-of-aging.com/timelines/hooke-history-cell-discovery.php

13

Discovery of Bacteria

Antonie van Leeuwenhoek

Antonie van Leeuwenhoek, a Dutch merchant who has taught himself how to build a microscope, sends a letter to the Royal Society of London detailing the strange things he has found living inside his mouth. Van Leeuwenhoek explains how, after scraping plaque from his teeth and putting it under his microscope, he was able to see, "with great wonder," many microscopic organisms that he called "animalcules."

He writes that they were "very prettily a-moving" and that "the biggest sort had a very strong and swift motion, and shot through the water (spittle) like a pike does through water." The

Discovery of Bacteria

second sort, he says, "oft-times spun round like a top... and these were far more in number."

Leeuwenhoek also describes what he had discovered in the plaque of two old men who had never brushed their teeth in their lives. He observed "an unbelievably great company of living animalcules, a-swimming more nimbly than any I had ever seen up to this time. The biggest sort... bent their body into curves in going forwards... Moreover, the other animalcules were in such enormous numbers, that all the water... seemed to be alive."

Leeuwenhoek's letter is the first record of bacteria seen living inside the human body.

It's not the first time that Leeuwenhoek has shared his discoveries with the Royal Society. Ten years earlier, he had first sent the organization a letter detailing his observations of what mold, lice and even bee stings looked like under a microscope. But in 1676, Leeuwenhoek had described something that the members of the Royal Society didn't believe was possible—he insisted that he had discovered microscopic living things inside pond scum scraped from a lake near his home. They remained dubious until Leeuwenhoek convinced them to see for themselves. And they found that he had been right—there were countless one-cell organisms moving around in the water.

The scientists no longer doubted Leeuwenhoek's findings--not in 1677 when he became the first person to observe spermatozoa swimming in human sperm, nor in 1682, when he described the nucleus inside the red blood cells of fish, and certainly not the following year when he shared his story about the strange tiny creatures he found inside his mouth.

Roughly 100 years later, in 1877, the Royal Society established the Leeuwenhoek Medal, awarded each decade to the person judged to have made the most significant contributions to the field of microbiology. One of the first winners was Louis Pasteur, who not only demonstrated that fermentation was caused by the growth of bacteria, but also discovered that a process that became known as "pasteurization" could kill bacteria.

It wasn't until 1876—200 years after Leeuwenhoek first saw bacteria "very prettily a-moving" under his microscope—that a German doctor named Robert Koch proved that those little organisms can causes disease.

http://www.healthcentral.com/dailydose/cf/2013/09/16/discovery_of_bacteria_in_humans_sept_17_1683

14

THE OCEAN CONTROLS THE WEATHER

The ocean can warm or cool the air in a number of different ways. For example, when the air is at a lower temperature than seawater, the ocean transfers heat to the lower atmosphere, which becomes less dense as the heat causes molecules in the air to move farther apart. As a result, a low-pressure air mass forms over that part of the ocean. (Conversely, cool or cold waters lead to the formation of high-pressure air masses as air molecules move closer together.) Because air always flows from areas of higher pressure to those of lower pressure, winds are diverted toward the low-pressure area.

Among winds that are affected by such pressure changes are the jet streams, bands of fast-moving, high-altitude air currents. Jet streams supply energy to developing storms at lower altitudes and then influence their movement. In this way, the ocean alters the direction of storm tracks. Some storms even reverse direction as the result of ocean-influenced air-pressure changes.

The ocean's currents make it possible for these weather effects to be widely distributed. Some currents carry warm water from tropical and subtropical regions toward the poles, while other currents move cool water in the opposite direction. The Gulf Stream is a current that transports warm water across the NorthAtlantic Ocean from Florida toward Europe. Before reaching Europe, the Gulf Stream breaks up into several other currents, one of which flows to the British Isles and Norway. The heat carried in this current warms the winds that blow over these regions, helping to keep winters there from becoming bitterly cold.

In this way, the ocean's circulation compensates somewhat for the sun's unequal heating of the Earth, in which the tropics receive more energy from the sun than the poles. Were it not for the moderating effects of ocean currents on air temperatures, the tropics would be much hotter than they are and the polar regions even colder.

Besides transferring heat to the atmosphere, the ocean also adds water to the air through evaporation. When the sun's heat causes surface water to evaporate, warm water vapor rises into the atmosphere. As the water vapor rises higher, it cools into tiny water droplets and ice crystals, which collect together to form large clouds. The clouds soon return their moisture to the surface as rain, snow, sleet, or hail. Most evaporation occurs in the warm waters of the tropics and subtropics, providing moisture for tropical storms.

Virtually all rain comes from the evaporation of seawater. Though this may seem surprising, it makes sense when one considers that about 97 percent of all water on Earth is in the ocean. The Earth's water cycle, or hydrologic cycle, consists largely of the never-ending circulation of water from the ocean to the atmosphere and then back to the ocean.

http://science.howstuffworks.com/how-the-ocean-affects-climate-info1.htm

15

THE HELIOCENTRIC MODEL

Nicolaus Copernicus

The Sun is the Centre of the Universe

The heliocentric model is a theory that places the Sun as the center of the universe, and the planets orbiting around it. The heliocentric model replaced geocentrism, which is the belief that the Earth is the center of the universe. The geocentric model was the prevailing theory in Ancient Greece, throughout Europe, and other parts of the world for centuries. It was not until the 16th century that the heliocentric model began to gain popularity because technology progressed enough to gain more evidence in its favor. Although heliocentrism did not gain popularity until the 1500's, the idea had existed for centuries throughout the world. In fact, Aristarchus of Samos – Samos was an island near Turkey – developed a form of the

The Heliocentric Model

heliocentric model as early as approximately 200 B.C. Other ancient civilizations held the same beliefs including various Muslim scholars in the 11th century who built on Aristarchus' work and European scholars in Medieval Europe.

In the 16th century, the astronomer Nicolaus Copernicus devised his version of the heliocentric model. Like other before him, Copernicus built on Atistarchus' work, mentioning the Greek astronomer in his notes. Copernicus' theory became so well known that when most people discuss the heliocentric theory today, they are referring to Copernicus' model. Copernicus published his theory in his book On the Revolutions of the Heavenly Bodies. Copernicus placed the Earth as the third planet from the Sun, and in his model, the Moon orbits the Earth not the Sun. Copernicus also hypothesized that the stars do not orbit the Earth; the Earth rotates, which makes the stars look like they have moved in the sky. Through the use of geometry, he was able to turn the heliocentric model from a philosophical hypothesis to a theory that did a very good job predicting the movement of the planets and other celestial bodies.

One problem facing the heliocentric model was that the Roman Catholic Church, a very powerful organization in Copernicus' time, considered it heretical. This may have been one of the reasons why Copernicus did not publish his theory until he was on his deathbed. After Copernicus died, the Roman Catholic Church worked even harder to suppress the heliocentric view. The Church arrested Galileo for promoting the heretical heliocentric model and kept him in house arrest for the last eight years of his life. Around the same time that

The Heliocentric Model

Galileo created his telescope, the astronomer Johannes Kepler was refining the heliocentric model and trying to prove it with calculations.

Although its progress was slow, the heliocentric model eventually replaced the geocentric model. As new evidence appeared though, some began to question whether the Sun was actually the center of the universe. The Sun is not the geometric center of the planets' orbits, and the center of gravity of the Solar System is not quite at the center of the Sun. What this means is that although children are taught in schools that heliocentrism is the correct model of the universe, astronomers use either view of the universe depending on what they are studying, and what theory makes their calculations easier.

http://www.universetoday.com/33113/heliocentric-model/

16

Satellites of Jupiter

Galileo Galilei

Jupiter has a large number of satellites. Of these, four are comparable to the Earth's Moon in size; the rest are orders of magnitude smaller. When Jupiter is at opposition and closest to the Earth, the stellar magnitude of its four large moons is between 5 and 6. This means that, were it not for the shielding brightness of Jupiter, these bodies would be visible with the naked eye. The aperture of the telescope used by Galileo in 1610 and its magnification thus brought these four "Galilean" satellites within his grasp.

But first Galileo had to make adjustments to the instruments. When viewing bodies that are very bright and very small, the optical defects of the telescope can be crippling. By trial and error Galileo learned to stop down the aperture of his instrument until he could begin to make useful observations. At the end of 1609, as he was finishing his series of observations of the Moon, Jupiter was at opposition and the brightest object in the evening sky (not counting the Moon). When he had made the new adjustment to his instrument, he turned his attention to Jupiter. On 7 January 1610 he observed the planet and saw what he thought were three fixed stars near it, strung out on a line through the planet. This formation caught his attention, and he returned to it the following evening.

Galileo's expectation was that Jupiter, which was then in its retrograde loop, would have moved from east to west and had left the three little stars behind. Instead, he saw all three stars to the west of Jupiter. It appeared as though Jupiter had not moved to the west but rather to the east. This was an anomaly, and Galileo returned to this formation again and again. Over the next week he found out several things. First, the little stars never left Jupiter; they appeared to be carried along with the planet. Second, as they were carried along, they changed their position with respect to each other and Jupiter. Third, there were not three but four of these little stars. By the 15th of January he had figured it out: these were not fixed stars but rather planetary bodies that revolved around Jupiter. Jupiter had four moons. His book, SidereusNuncius, in which his discovery was described, came off the press in Venice in the middle of March 1610 and made Galileo famous.

http://galileo.rice.edu/sci/observations/jupiter_satellites.html

17

INFRARED AND ULTRAVIOLET

Frederick Herschel

Frederick Herschel was born in Hanover, Germany, in 1738. As a young man, he grew into a gifted musician and astronomer. It was Herschel who discovered the planet Uranus in 1781, the first new planet discovered in almost 2,000 years. In late 1799 Herschel began a study of solar light. He often used color filters to isolate parts of the light spectrum for these studies and noted that some filters grew hotter than others. Curious about this heat in solar radiation, Herschel wondered if some colors naturally carried more heat than others.

To test this idea, Herschel built a large prism. In a darkened room, he projected the prism's rainbow light spectrum onto the far wall and carefully measured the temperature inside each of these separate col ored light beams.

Herschel was surprised to find that the temperature rose steadily from violet (coolest) to a maximum in the band of red light. On a sudden impulse, Herschel placed a thermometer in the dark space right next to the band of red light (just be yond the light spectrum). This thermometer should have stayed cool. It was not in any direct light. But it didn't.

This thermometer registered the most heat of all.

Herschel was amazed. He guessed that the sun radiated heat waves along with light waves and that these invisible heat rays refract slightly less while traveling through a prism than do light rays. Over the course of several weeks, he tested heat rays and found that they refracted, reflected, bent, etc., exactly like light. Because they appeared below redlight, Herschel named them infrared (mean ing below red).

Johann Ritter was born in 1776 in Ger many and became a natural science philosopher. His central beliefs were that there was unity and symmetry in nature and that all natural forces could be traced back to one prime force, Urkraft. In 1801, Ritter read about Herschel's discovery of infrared radiation. Ritter had worked on sunlight's effect on chemical reactions and with electrochemistry (the effect of electrical currents on chemicals and on chemical reactions). During this work he had tested light's effect on silver chloride and knew that exposure to light turned this chemical from white to black. (This discovery later became the basis for photography.)

Ritter decided to duplicate Herschel's experiment, but to see if all colors darkened silver chloride at the same rate. He coated strips of paper with silver chloride. In a dark room he repeated Herschel's set up. But instead of measuring temperature in each color of the rainbow spectrum projected on wall, Ritter timed how long it took for strips of silver chloride paper to turn black in each color of the spectrum.

He found that red hardly turned the paper black at all. He also found that violet darkened paper the fastest. Again mimicking Herschel's experiment, Ritter placed a silver chloride strip in the

dark area just be yond the band of violet light. This strip blackened the fastest of all! Even though this strip was not exposed to visible light, some radiation had acted on the chemicals to turn them black.

Ritter had discovered radiation beyond violet (ultra violet) just as Herschel had discovered that radiation existed below the red end of the visible spectrum (infrared).

http://www.dadazi.net/new/diff/100science/book03/scbk03-06.html

18

Relativity

Albert Einstein

Albert Einstein's theory of relativity is actually two separate theories: his special theory of relativity, postulated in the 1905 paper, The Electrodynamics of Moving Bodies and histheory of general relativity, an expansion of the earlier theory, published as The Foundation of the General Theory of Relativity in 1916. Einstein sought to explain situations in which Newtonian physics might fail to deal successfully with phenomena, and in so doing proposed revolutionary changes in human concepts of time, space, and gravity.

The special theory of relativity was based on two main postulates: first, that the speed of light is constant for all observers; and second, that observers moving at constant speeds should be subject to the same physical laws. Following this logic, Einstein theorized that time must change according to the speed of a moving object relative to the frame of reference of an observer. Scientists have tested this theory through experimentation - proving, for example, that an atomic clock ticks more slowly when traveling at a high speed than it does when it is not moving. The essence of Einstein's paper was that both space andtime are relative (rather than absolute), which was said to hold true in a special case, the absence of a gravitational field. Relativity was a stunning concept at the time; scientists all over the world debated the veracity of Einstein's famous equation, $E=mc2$, which implied that matter and energy were equivalent and, more specifically, that a single particle of matter could be converted into a huge quantity of energy. However, since the special theory of relativity only held true in the absence of a gravitational field, Einstein strove for 11 more years to work gravity into his equations and discover how relativity might work generally as well.

According to the Theory of General Relativity, matter causes space to curve. It is posited that gravitation is not a force, as understood by Newtonian physics, but a curved field (an area of space under the influence of a force) in the space-time continuum that is actually created by the presence of mass. According to Einstein, that theory could be tested by measuring the deflection of starlight traveling near the sun; he correctly asserted that light deflection would be twice that expected by Newton's laws. This theory also explained why the light from stars in a strong gravitational field was closer to the red end of the spectrumthan those in a weaker one.

For the final thirty years of his life, Einstein attempted to find a unified field theory , in which the properties of all matter and energy could be expressed in a single equation. His search was confounded by quantum theory 's uncertainty principle , which stated that the movement of a single particle could never be accurately measured, because speed and position could not be simultaneously assessed with any degree of assurance. Although he was unable to find the comprehensive theory that he sought, Einstein's pioneering work has allowed countless other scientists to carry on the quest for what some have called "the holy grail of physicists."

http://whatis.techtarget.com/definition/theory-of-relativity

19

The Doppler Effect

Christian Doppler

When we are moving, or a source producing a sound is moving, we hear things differently. You may have noticed that a train whistle gets lower as it passes you. The whistle is not changing pitch, but you are hearing a change. This principle is known as the Doppler effect. The Doppler effect is named after the Austrian physicist, Christian Johann Doppler, who discovered it.

Doppler claimed that if a sound is getting closer to you, either because its source is approaching you or because you are going towards the source, the sound will seem higher than it really is. If you are heading away from a source or it is going away from you, he believed the sound would seem lower than its actual pitch. To test his theory, scientists hired 15 trumpeters to play on a moving train. As the train passed by them, they heard a drop in pitch, just like Doppler predicted.

The Doppler Effect

The Doppler effect happens because distance affects the amount of time it takes you to hear the sound. Imagine you are playing in the park and your friend rolls a ball to you. The ball would reach you sooner if you walked towards it and later if you moved away from it. The same is true for sound. Remember that frequency is wavelengths per time. If you hear a frequency in a shorter amount of time, it seems like you are hearing a higher frequency. For example, say you heard a sound that had 50 wavelengths by the time it reached you, it would have taken it 5 seconds to reach you. The frequency of that sound is 50 divided by 5, or 10 Hertz. Imagine you heard the same sound, but this time you were moving towards its source and it only took 2 seconds for 50 wavelengths to reach you. Now the frequency you hear is 50 divided by 2, or 25 Hertz. The frequency seemed higher because you were moving. If you were not moving, after 2 seconds, only 20 wavelengths would have reached you and the frequency would still sound like 10 Hertz.

The opposite happens when the distance between you and a source of sound widens. Now it takes longer for you to hear a certain amount of wavelengths. Therefore, the frequency seems lower. The Doppler effect makes a pitch appear to change when you, or the source, are in motion.

http://www.ndt-ed.org/EducationResources/HighSchool/Sound/dopplereffect.htm

20

The Heart and the Circulatory System

William Harvey

William Harvey was born in 1578 in Folkestone, England. The eldest of seven sons, Harvey received a Bachelor of Arts degree from Cambridge in 1597. He then studied medicine at the University of Padua, receiving his doctorate in 1602. By all measures, Harvey was successful. After he finished his studies at Padua, he returned to England and set up practice. He then married Elizabeth Brown, daughter of the court physician to Queen Elizabeth I and King James I. This put in him in position to be noticed by the aristocracy, and Harvey quickly moved up the ladder. Eventually, he became court physician to both King James I and King Charles I.

The Heart and The Circulatory System

While acting as court physician, Harvey was able to conduct his research in human biology and physiology. Harvey focused much of his research on the mechanics of blood flow in the human body. Most physicians of the time felt that the lungs were responsible for moving the blood around throughout the body. Harvey questioned these beliefs and his questions directed his life-long scientific investigations.

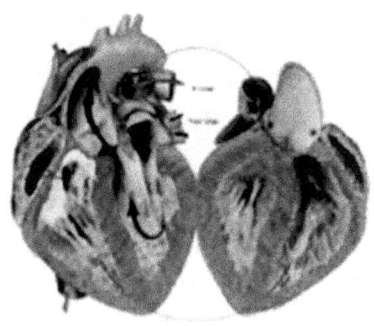

Image courtesy of Carolina Biological Supply/Access Excellence

Harvey's experiments involved both direct dissection and physiological experiments on animals. His observations of dissected hearts showed that the valves in the heart allowed blood to flow in only one direction. Direct observation of the heartbeat of living animals showed that the ventricles contracted together, dispelling Galen's theory that blood was forced from one ventricle to the other. Dissection of the septum of the heart showed that it contained arteries and veins, not perforations. When Harvey removed the beating heart from a living animal, it continued to beat, thus acting as a pump, not a sucking organ. Harvey also used mathematical data to prove that the blood was not being consumed. Removal of the blood from human cadavers showed that the heart could hold roughly two ounces of blood. By calculating the number of heartbeats in a day and multiplying this by two ounces, he showed that the amount of blood pump far exceeded the amount that the body could possibly make. He based this figure on how much food and liquids a person could consume. To Harvey, this showed that the teaching by Galen that the blood was being consumed by the organs of the body was false. Blood had to be flowing through a 'closed circuit' instead. Even though he lacked a microscope, Harvey theorized that the arteries and veins were connected to each other by capillaries, which would later be discovered by Marcello Malpighi some years after Harvey's death.

Harvey did not let the beliefs of Galen concerning the role of natural, vital, and animal spirits and their effects on physiology affect his objectivity. Instead, Harvey asked simple, pointed questions, the types of questions that even today are the hallmark of good scientific research. Harvey asked such questions as why did both the lungs and the heart move if only the lungs were responsible for causing circulation of blood? Why should, as Galen suggested, structurally similar parts of the heart have very different functions? Why did 'nutritive' blood appear so similar to 'vital' blood? These, and other, questions gave Harvey his focus.

Harvey's lecture notes show that he believed in the role of the heart in circulation of blood through a closed system as early as 1615. Yet he waited 13 years, until 1628, to publish his findings in his work Exercitatioanatomica de motucordisetsanguinis in animalibus or On the Movement of the Heart and Blood in Animals. Why did he wait so long? Galenism, or the study and practice of medicine as originally taught by Galen, was almost sacred at the time Harvey lived. No one dared to challenge the teachings of Galen. Like most physicians of his day, William Harvey, was trained in the ways of Galen. Conformation was not only the norm, but was also the key to success. To rebel against the teachings of Galen could quickly end the career of any physician. Perhaps this is why he waited.

Harvey's hesitation proved well-founded. After his work was published, many physicians and scientists rejected him and his findings. Using different assumptions of the amount of blood contained in the heart, scientists argued that the blood could indeed be consumed. Controversy raged for a full twenty years after publication of "On the Movement of the Heart and Blood in Animals." Yet, with time, more and more physicians and researchers accepted Harvey's hypotheses.

Like all good research, Harvey's work raised more questions than it answered. For example, if blood was not consumed by organs, how did different parts of the body obtain nourishment? If the liver did not make blood from food, where did blood originate? These questions, and others like them, directed the research of many investigations for many years to come. Medical practice in Harvey's time, however, changed little. Even though the mechanics of blood flow were understood now, the understanding of the causes of many diseases were still bathed in the mystery of spirits. In fact, the practices of bleeding, lancing, and leeching increased in the years following Harvey's work. On the positive side, medicine did make some advances, for it was during the seventeenth century that administering medicine through intravenous injections came into practice.

William Harvey's classic work became the foundation for all modern research on the heart and cardiovascular medicine. It has been said that Harvey's proof "of the continuous circulation of the blood within a contained system was the seventeenth century's most significant achievement in physiology and medicine." Further, his work is considered to be one of the most important contributions in the history of medicine. Without the understanding of the circulatory system made possible by Harvey's pioneering work, the medical miracles that we think are commonplace would be impossible.

http://biology.about.com/library/organs/blcircsystem2.htm

21

Who Discovered Electricity?

Nancy Atkinson

Electricity is a form of energy and it occurs in nature, so it was not "invented." As to who discovered it, many misconceptions abound. Some give credit to Benjamin Franklin for discovering electricity, but his experiments only helped establish the connection between lightning and electricity, nothing more.

The truth about the discovery of electricity is a bit more complex than a man flying his kite. It actually goes back more than two thousand years.

In about 600 BC, the Ancient Greeks discovered that rubbing fur on amber (fossilized tree resin) caused an attraction between the two – and so what the Greeks discovered was actually static electricity. Additionally, researchers and archeologists in the 1930's discovered pots with sheets of copper inside that they believe may have been ancient batteries meant to produce light at ancient Roman sites. Similar devices were found in archeological digs near Baghdad meaning ancient Persians may have also used an early form of batteries.

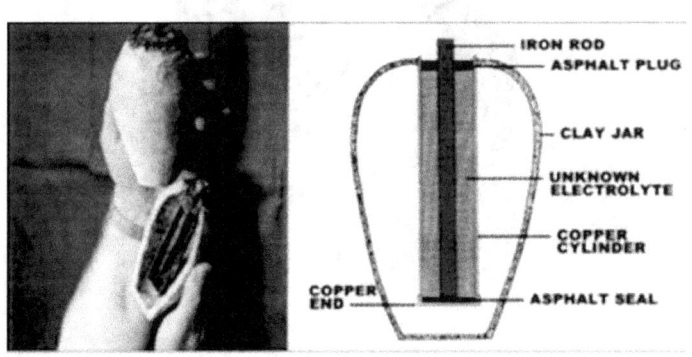

A replica and diagram of one of the ancient electric cells (batteries) found near Bagdad.

But by the 17th century, many electricity-related discoveries had been made, such as the invention of an early electrostatic generator, the differentiation between positive and negative currents, and the classification of materials as conductors or insulators.

In the year 1600, English physician William Gilbert used the Latin word "electricus" to describe the force that certain substances exert when rubbed against each other. A few years later another English scientist, Thomas Browne, wrote several books and he used the word "electricity" to describe his investigations based on Gilbert's work.

Benjamin Franklin

In 1752, Ben Franklin conducted his experiment with a kite, a key, and a storm. This simply proved that lightning and tiny electric sparks were the same thing.

Who Discovered Electricity?

Italian physicist Alessandro Volta discovered that particular chemical reactions could produce electricity, and in 1800 he constructed the voltaic pile (an early electric battery) that produced a steady electric current, and so he was the first person to create a steady flow of electrical charge. Volta also created the first transmission of electricity by linking positively-charged and negatively-charged connectors and driving an electrical charge, or voltage, through them.

In 1831 electricity became viable for use in technology when Michael Faraday created the electric dynamo (a crude power generator), which solved the problem of generating electric current in an ongoing and practical way. Faraday's rather crude invention used a magnet that was moved inside a coil of copper wire, creating a tiny electric current that flowed through the wire. This opened the door to American Thomas Edison and British scientist Joseph Swan who each invented the incandescent filament light bulb in their respective countries in about 1878. Previously, light bulbs had been invented by others, but the incandescent bulb was the first practical bulb that would light for hours on end.

Replica of Thomas Edison's first lightbulb. Credit: National Park Service.

Who Discovered Electricity?

Swan and Edison later set up a joint company to produce the first practical filament lamp, and Edison used his direct-current system (DC) to provide power to illuminate the first New York electric street lamps in September 1882.

Later in the 1800's and early 1900's Serbian American engineer, inventor, and all around electrical wizard Nikola Tesla became an important contributor to the birth of commercial electricity. He worked with Edison and later had many revolutionary developments in electromagnetism, and had competing patents with Marconi for the invention of radio. He is well known for his work with alternating current (AC), AC motors, and the polyphase distribution system. Later, American inventor and industrialist George Westinghouse purchased and developed Tesla's patented motor for generating alternating current, and the work of Westinghouse, Tesla and others gradually convinced American society that the future of electricity lay with AC rather than DC. Others who worked to bring the use of electricity to where it is today include Scottish inventor James Watt, Andre Ampere, a French mathematician, and German mathematician and physicist George Ohm.

And so, it was not just one person who discovered electricity. While the concept of electricity was known for thousands of years, when it came time to develop it commercially and scientifically, there were several great minds working on the problem at the same time.

http://www.universetoday.com/82402/who-discovered-electricity/#ixzz3065GQWRP

22

Theory of Evolution by Charles Darwin

Stephen Montgomery

Natural Selection

Natural selection is Darwin's most famous theory; it states that evolutionary change comes through the production of variation in each generation and differential survival of individuals with different combinations of these variable characters. Individuals with characteristics which increase their probability of survival will have more opportunities to reproduce and their offspring will also benefit from the heritable, advantageous character. So over time these variants will spread through the population.

Natural Selection In The Evolutionary Framework:

For natural selection to work, it has to occur along with a bunch of other things. Historians and biologists who have analysed Darwin's work, for example Ernst Mayr, have identified fivetheories which Darwin outlined in On the Origin of Species, and which work together to bring about evolution.

Darwin's five theories were:

i. Evolution: species come and go through time, while they exist they change.

ii. Common descent: organisms are descended from one, or several common ancestors and have diversified from this original stock

iii. Species multiply: the diversification of life involves populations of one species diverging until they become two separate species; this has probably occurred billions of times on earth!

iv. Gradualism: evolutionary change occurs through incremental small changes within populations; new species are not created suddenly.

v. Natural selection: evolutionary change occurs through variation between individuals; some variants give the individual an extra survival probability.

Darwin considered all these theories as parts of one grand idea; they all occur together. Scientists however took a while to see this; they weren't accepted as a package until the modern synthesis of the 1930/40s. Before then scientists would favour some ideas but propose alternatives to fill in the gaps, natural selection was one of the least popular, to find out why click here. Eventually, as more evidence accumulated and these different ideas were tested it became clear that Darwin was right all along!

How Does Natural Selection Work?

Natural selection was Darwin's most novel and revolutionary idea, but in truth (like all the best ideas) it is very simple. Despite its simplicity, since the publication of the theory right up until today, it has widely been misunderstood. Ernst Mayr, in his book One Long Argument (1991) provides a useful way of breaking down the process into just five facts and three inferences, or conclusions, drawn from the five facts; they can be linked in a flow diagram:

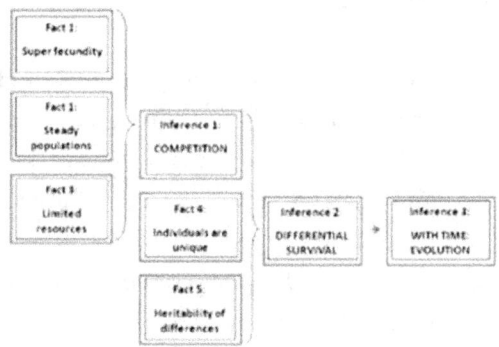

Theory of Evolution by Charles Darwin

Figure: modified from One Long Argument by Ernst Mayr (1991)

The first inference is drawn from three facts which Darwin observed in the natural world around him. He saw that organisms produce more offspring than is required to replace themselves, so population sizes should increase rapidly (think about the number of frogspawn laid each year, or how many eggs a spider lays). That's fact one: a fancy word for this over-reproduction is 'super fecundity'. However Darwin saw for himself, and confirmed his observation with others, that population numbers tend to stay at about the same level (you don't see a doubling of the number of frogs or mice in your garden each year do you?): that's fact two. What accounts for this disparity? Darwin found the answer with another fact: resources, such as food, water or places to sleep or mate, are limited. A major influence on Darwin observing this fact was his reading the work of Thomas Malthus who published a paper stating that the human population was increasing at a rapid pace and would soon run out of food, water and space. These are three simple facts which Darwin put together to draw a simple conclusion: individuals compete with each other for scarce resources.

Next, Darwin made two other observations about individuals. First he had come to the conclusion through his work on the H.M.S. Beagle, when he was working on barnacles and later pigeons, that individuals are unique and that individuals vary in almost every aspect: that's fact four, and you only need to take a cursory glance round a group of people to see that it is true! Finally fact five: Darwin had taken to breeding pigeons to investigate variability further. He performed many crosses between different breeds of fancy pigeons to look at whether their offspring had the same variations. He also collected lots of observations from various animal and plant breeders to help him draw out the conclusion that these individual differences are heritable: they are passed on from parent to offspring.

The next two inferences demonstrate Darwin's genius. Darwin could see that if individuals must compete, and if they are all unique, some individuals will have variations which give them a survival boost so they will have more opportunity to reproduce and leave a greater number of offspring. These offspring will inherit the variations which made their parents successful, so they too will have an advantage. Over time these successful variantions will spread through the population – the population will change: that is evolution! Simple, isn't it?

The Evidence

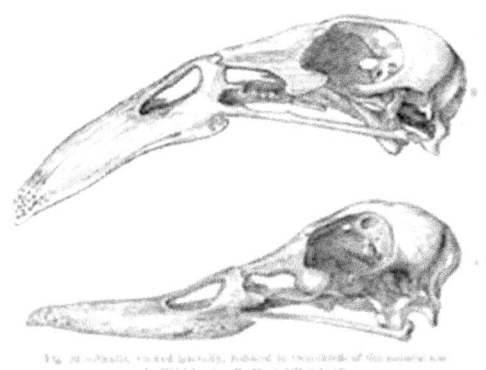

Theory of Evolution by Charles Darwin

Darwin himself wrote 'an unverified hypothesis is of little or no value'. To verify his 'hypothesis' Darwin collected a vast number of facts from a wide range of fields. He assembled reports from other naturalists, as well as from his own work and observation, to support his five facts. His greatest challenge perhaps, was to convince people that species really are variable and that this variation is suitable for natural selection to act. Darwin chose to demonstrate this using artificial selection and production of various breeds of domestic animals and plants as an analogy for natural selection. You can read more about how he did this by clicking here.

Darwin added to his bulk of evidence throughout his lifetime, for example with studies on humans. Since 1859 the scientific community has been busy testing his theories, and alternatives, to see what best holds up. The wealth and diversity of evidence is now vast and includes evidence from the DNA record, fossil record, and from case studies section.

http://darwin200.christs.cam.ac.uk/pages/index.php?page_id=d3

23

THE DISCOVERY OF RADIO WAVES

Heinrich Hertz

Heinrich Hertz
(1857-1894)

German physicist Heinrich Hertz discovered radio waves, a milestone widely seen as confirmation of James Clerk Maxwell's electromagnetic theory and which paved the way for numerous advances in communication technology. Born in Hamburg on February 22, 1857, Hertz was the eldest of five children. His mother was Elizabeth Pfefferkorn

Hertz and his father was Gustav Hertz, a respected lawyer who would later become a legislator. In his youth Heinrich displayed an interest in building things, and as a teenager he constructed a spectroscope and a galvanometer that were so well designed that Hertz used them throughout his college years. Initially Hertz planned a career in engineering, but after a year of employment at the public works office in Frankfurt, a summer of classes at the Polytechnic in Dresden, a year of military service in Berlin, and a brief stint in the engineering department at the University of Munich, he finally decided to pursue the subject that most deeply interested him: science.

Throughout his life Hertz read works on science and carried out experiments as a hobby. But once he decided that science was to be his career, he applied himself to these tasks with even greater enthusiasm. In the winter of 1877, he studied various scientific treatises, and the following spring, he gained some laboratory experience by working with Gustav von Jolly. Subsequently he enrolled at the University of Berlin, where he was privileged to study under the great German physicist Hermann von Helmholtz. With Helmholtz's encouragement, Hertz resolved to compete for a research prize to be awarded to the student best able to determine whether or not electricity moved with inertia. Hertz began a series of experiments into the matter, and this mode of learning seemed to suit him. He confided in a letter sent to his family during that time, "I cannot tell you how much more satisfaction it gives me to gain knowledge for myself and for others directly from nature, rather than merely learning from others and for myself alone."

The Discovery of Radio Waves

In August of 1879, Hertz won the prize for his evidence demonstrating that electricity had no inertia. Another prize problem was soon proposed by Helmholtz, who wanted students to attempt to prove which of the theories of electromagnetic phenomena then circulating was correct. Interestingly, Hertz did not choose to compete for this prize, but years later would be the first person to successfully provide the kind of definitive evidence that Helmholtz sought. At the time, Hertz instead embarked on a study of induction produced by rotating spheres. His work in this area helped him earn his doctorate degree ahead of schedule, in 1880, magna cum laude.

Hertz's first academic post was as lecturer of theoretical physics at the University of Kiel, but due to his dissatisfaction there he accepted a position at the Karlsruhe Polytechnic in 1885. It was at Karlsruhe, where he remained until he received an appointment as physics professor at the University of Bonn in 1889, that Hertz carried out his most important work. In 1886, Hertz began experimenting with sparks emitted across a gap in a short metal loop attached to an induction coil. He soon built a similar apparatus, but without the induction coil, to act as a detector. When the induction coil connected to the first loop (the transmitter) produced a high voltage discharge, a spark jumped across the gap, sending out a signal that Hertz detected as a weaker spark across the gap in the receiving apparatus, which he placed nearby. To determine the nature of the signals that he was able to transmit and receive, Hertz developed a number of innovative experiments.

By measuring side sparks that formed around the primary spark and varying the position of the detector, Hertz was able to determine that the signal exhibited a wave pattern, and to ascertain its wavelength. Then, by using a rotating mirror, he found the frequency of the invisible waves, which enabled him to calculate their velocity. Amazingly, the waves were moving at the speed of light. Thus, it appeared to Hertz that he had discovered a previously unknown form of electromagnetic radiation, and in the process confirmed James Clerk Maxwell's theory of electromagnetism. To further prove that this was indeed the case, Hertz continued his experiments exploring the behavior of the invisible waves. He discovered that they traveled in straight lines and could be focused, diffracted, refracted and polarized. Hertz announced his initial discovery in late 1887 in his treatise "On Electromagnetic Effects Produced by Electrical Disturbances in Insulators", which he sent to the Berlin Academy. He later published additional details following the series of experiments he carried out in 1888. For a time the waves he discovered were commonly referred to as Hertzian waves, but today they are known as radio waves.

In addition to his radio wave breakthrough, Hertz is notable for the discovery of the photoelectric effect, which occurred while he was investigating electromagnetic waves. Because of some difficulty in detecting the small spark produced in his receiving apparatus, Hertz sometimes placed the receiver in a dark case. This, he found, affected the maximum length of the spark, which was smaller than when he did not use the case. With further research into the phenomenon, Hertz discovered that the spark produced was stronger if it was exposed to ultraviolet light. Though he did not attempt to explain this fact, others, including J.J. Thomson and Albert Einstein, would soon realize its importance. The phenomenon of electrons being released from a material when it absorbs radiant energy, which was the cause of the stronger sparks observed by Hertz when ultraviolet radiation was used, would come to be known as the photoelectric effect.

After 1889, when Hertz was teaching at the University of Bonn, he studied electrical discharges in rarefied gases and spent a significant amount of time composing his Principles of Mechanics. Unfortunately, he never saw the work published due to his premature death associated with blood poisoning on New Year's Day 1894. Only 37 years old at the time, Hertz also never lived to see the tremendous impact the discovery of radio waves would have on the world in the 20th century.

http://www.magnet.fsu.edu/education/tutorials/pioneers/hertz.html

24

Splitting the Atom

Ernest Rutherford

Rutherford was born on 30 August, 1871 at Brightwater, just south of Nelson, New Zealand. His father was the son of a Scots colonist from Dundee, and his mother an Englishwoman from Sussex. Rutherford attended schools close to his home before entering Nelson College in 1887.

In 1889 Rutherford gained a junior scholarship and left for Canterbury College, Christchurch, at the University of New Zealand. He graduated in 1893 with a Double First in Physics and Mathematics, only the second person to do so in the history of the University. Rutherford received an M.A. for this work, and began research towards his B.Sc., which he received the following year.

In 1894 Rutherford was awarded an 1851 Exhibition Scholarship that allowed him to travel abroad to do research in physics. He chose to go to the University of Cambridge, which had just started to accept research students from overseas. Rutherford travelled to England and entered Trinity College, leaving his fiancée Mary Newton in New Zealand.

Rutherford had been investigating ways of sending and detecting 'Hertzian waves', and had already managed to send these radio signals over the length of his laboratory. In Cambridge he worked on improving the range and could soon send signals over half a mile, the record for a long-distance transmission at the time.

In November 1895 Wilhelm Conrad Röntgen discovered X-rays in Germany. Laboratories all over the world quickly reproduced his results, and in the Cavendish J.J. Thomson realised that X-rays made gases conduct by 'splitting up', or ionising, their molecules. Rutherford and Thomson investigated this and found the conductivity vanished if a current was passed through the gas, explaining that 'the current destroys and the rays produce the structure which gives conductivity to the gas'.

Meanwhile, in Paris, Antoine Henri Becquerel had discovered the radioactivity of uranium salts. Rutherford investigated the rays given off by uranium and found that he could split them into three types using a magnetic field. The first type, which Rutherford named alpha-rays, were positively charged and were deflected by the magnetic field. The second type, beta-rays, were negatively charged and were deflected in the opposite direction. The third type, gamma-rays, werenot deflected at all.

Rutherford's experiments in the Cavendish showed that the gamma-rays were a type of X-ray, and that the alpha- and beta-rays were tiny particles of matter. Rutherford showed that the beta-particles were electrons, but the alpha-particles were something new.

In 1898 Rutherford was offered the professorship of Physics at McGill University in Montreal, Canada. Although Rutherford was happy at Cambridge, the McGill position paid a salary of £500 a year, enough to allow Rutherford to marry his fiancée in New Zealand. In September 1898 Rutherford left Cambridge for Montreal.

At McGill, Rutherford continued to study radioactivity. Working with R.B. Owens he found that thorium gave off a radioactive gas that could induce radioactivity in other substances.

In the summer of 1900 Rutherford visited New Zealand and married his fiancée, Mary Newton. The two of them returned to McGill in time for the autumn term, when Rutherford was joined by Frederick Soddy, a chemist from Oxford University.

Rutherford and Soddy investigated different radioactive metals. They found that radioactive materials break down into different elements, giving off energetic particles in the process. They could draw decay curves of this process, showing how substances lost their radioactivity over time. Together Rutherford and Soddy investigated the alpha-particles, and concluded by their mass and charge that they must be charged helium atoms. Rutherford explained his discoveries in a book, Radio-activity, which was published in 1904.

Rutherford was offered positions at several American universities but turned them down. Then, in 1906, he was invited to apply for the post of Langworthy Professor of Physics at the University of Manchester. Rutherford applied, and in January 1907 he learned that he had been appointed. Rutherford returned to England with his wife and young daughter.

In Manchester Rutherford met Hans Geiger, a young German already working in the department. Rutherford acquired some radium from Vienna, allowing Geiger to develop the Geiger Counter, a device that detects individual alpha-particles. By using the Geiger Counter precise rates of alpha particle emission could be measured for different substances.

At McGill Rutherford had concluded that alpha-particles were helium ions. In 1908 he demonstrated this with a clever piece of apparatus. Rutherford made an air-tight glass tube with very thin walls, and filled it with the radioactive radium emanation. Alpha-particles could penetrate the thin walls of the tube and were collected in a second tube. After some time these alpha-particles were compressed and a spark passed. The spectrum of the gas showed that the alpha-particles were helium ions.

Rutherford was honoured in 1908 when he received the Nobel Prize in Chemistry. He was surprised that his work was regarded as Chemistry, saying that he "had dealt with many different transformations with various time-periods, but the quickest he had met was his own transformation from a physicist to a chemist".

Rutherford began using alpha particles to investigate other elements. If a lot of alpha particles were fired into another substance, some would be deflected by small angles. In 1909 Rutherford suggested that one of Geiger's research students, Ernest Marsden, should see if he could detect similar deflections at very large angles. To Rutherford's surprise, Geiger and Marsden found that some alpha-particles were deflected almost completely! This experiment inspired Rutherford's theory of atomic nuclei, that all the positive charge and most of the mass of an atom must be contained in a tiny nucleus at the atom's centre. Rutherford presented this theory of the nuclear atom in 1911.

Just before the outbreak of the First World War Rutherford received a knighthood from King George V. Between 1914 and 1918 Rutherford worked on methods for finding enemy submarines, while continuing his own investigations of the nucleus. In 1919, a few months after the war ended, Rutherford showed that by firing alpha-particles into nitrogen gas a small amount of hydrogen could be produced. This was the first artificial disintegration of a nucleus.

In 1919 J.J. Thomson resigned as Cavendish Professor in Cambridge and the position was offered to Rutherford.

Splitting The Atom

Rutherford was at first a little reluctant, but accepted when he was sure that his appointment would not interfere with his and Thomson's friendship. In the autumn of 1919 Rutherford returned to the Cavendish as its Director.

Back in the Cavendish, Rutherford continued his experiments with alpha-particles. His recent experiments had shown that nuclei contained protons. In 1920 Rutherford suggested that nuclei might also contain a neutral particle of about the same mass as the proton. He and his assistant, James Chadwick, spent many years looking for these neutral particles.

In 1924 Rutherford was awarded the Order of Merit, one of the greatest honours anyone can receive. Rutherford felt that 'it was not an honour given to him as a person; he looked upon it as a recognition of the value and importance of science in the twentieth-century world'. Rutherford was further honoured in 1931 when the King made him First Baron Rutherford of Nelson, New Zealand, and Cambridge. Rutherford attended the House of Lords whenever scientific matters were discussed.

1932 was a great year for Rutherford and the Cavendish. Early in the year James Chadwick finally found proof of neutral particles inside the nucleus. Chadwick could produce these 'neutrons' by bombarding beryllium with alpha-particles. A few months later Ernest Cockcroft and J.D. Walton, who had been working on high voltage accelerators for several years, successfully 'split atoms' of lithium using accelerated protons. This was the first nuclear disintegration that was completely under the control of the experimenter. Both experiments gave firm support to Rutherford's belief that nuclei are made up of smaller particles.

Rutherford shaped the Cavendish into the world's most influential laboratory for nuclear physics. In his final years, many of the men Rutherford had supervised left him to head other important laboratories throughout the United Kingdom.

Rutherford had one daughter, Eileen, who married the physicist R.H. Fowler in 1921. Tragically, Eileen died on 23rd December 1930, a week after giving birth to her fourth child. Rutherford's grandchildren were one of the delights of his life, and he was always happy to talk about them.

On Thursday, 14th October 1937 Rutherford felt unwell and summoned his doctor the next day. An operation was performed that evening, but from Saturday afternoon onwards Rutherford's health deteriorated. He died on 19th October 1937, aged 66. Rutherford's ashes are buried in the nave of Westminster Abbey, near the tombs of Sir Isaac Newton and Lord Kelvin.

http://www-outreach.phy.cam.ac.uk/camphy/physicists/rutherford_prelim.htm

25

DISCOVERY OF DNA STRUCTURE AND FUNCTION: WATSON AND CRICK

Leslie A. Pray

The landmark ideas of Watson and Crick relied heavily on the work of other scientists. What did the duo actually discover?

Many people believe that American biologist James Watson and English physicist Francis Crick discovered DNA in the 1950s. In reality, this is not the case. Rather, DNA was first identified in the late 1860s by Swiss chemist Friedrich Miescher. Then, in the decades following Miescher's discovery, other scientists--notably, Phoebus Levene and Erwin Chargaff--carried out a series of research efforts that revealed additional details about the DNA molecule, including its primary chemical components and the ways in which they joined with one another. Without the scientific foundation provided by these pioneers, Watson and Crick may never have reached their groundbreaking conclusion of 1953: that the DNA molecule exists in the form of a three-dimensional double helix.

The First Piece of the Puzzle: Miescher Discovers DNA

Although few people realize it, 1869 was a landmark year in genetic research, because it was the year in which Swiss physiological chemist Friedrich Miescher first identified what he called "nuclein" inside the nuclei of human white blood cells. (The term "nuclein" was later changed to "nucleic acid" and eventually to "deoxyribonucleic acid," or "DNA.") Miescher's plan was to isolate and characterize not the nuclein (which nobody at that time realized existed) but instead the protein components of leukocytes (white blood cells). Miescher thus made arrangements for a local surgical clinic to send him used, pus-coated patient bandages; once he received the bandages, he planned to wash them, filter out the leukocytes, and extract and identify the various proteins within the white blood cells. But when he came across a substance from the cell nuclei that had chemical properties unlike any protein, including a much higher phosphorous content and resistance to proteolysis (protein digestion), Miescher realized that he had discovered a new substance (Dahm, 2008). Sensing the importance of his findings, Miescher wrote, "It seems probable to me that a whole family of such slightly varying phosphorous-containing substances will appear, as a group of nucleins, equivalent to proteins" (Wolf, 2003).

Discovery of DNA Structure and Function

More than 50 years passed before the significance of Miescher's discovery of nucleic acids was widely appreciated by the scientific community. For instance, in a 1971 essay on the history of nucleic acid research, Erwin Chargaff noted that in a 1961 historical account of nineteenth-century science, Charles Darwin was mentioned 31 times, Thomas Huxley 14 times, but Miescher not even once. This omission is all the more remarkable given that, as Chargaff also noted, Miescher's discovery of nucleic acids was unique among the discoveries of the four major cellular components (i.e., proteins, lipids, polysaccharides, and nucleic acids) in that it could be "dated precisely... [to] one man, one place, one date."

Laying the Groundwork: Levene Investigates the Structure of DNA

Meanwhile, even as Miescher's name fell into obscurity by the twentieth century, other scientists continued to investigate the chemical nature of the molecule formerly known as nuclein. One of these other scientists was Russian biochemist Phoebus Levene. A physician turned chemist, Levene was a prolific researcher, publishing more than 700 papers on the chemistry of biological molecules over the course of his career. Levene is credited with many firsts. For instance, he was the first to discover the order of the three major components of a single nucleotide (phosphate-sugar-base); the first to discover the carbohydrate component of RNA (ribose); the first to discover the carbohydrate component of DNA (deoxyribose); and the first to correctly identify the way RNA and DNA molecules are put together.

Discovery of DNA Structure and Function

During the early years of Levene's career, neither Levene nor any other scientist of the time knew how the individual nucleotide components of DNA were arranged in space; discovery of the sugar-phosphate backbone of the DNA molecule was still years away. The large number of molecular groups made available for binding by each nucleotide component meant that there were numerous alternate ways that the components could combine. Several scientists put forth suggestions for how this might occur, but it was Levene's "polynucleotide" model that proved to be the correct one. Based upon years of work using hydrolysis to break down and analyze yeast nucleic acids, Levene proposed that nucleic acids were composed of a series of nucleotides, and that each nucleotide was in turn composed of just one of four nitrogen-containing bases, a sugar molecule, and a phosphate group. Levene made his initial proposal in 1919, discrediting other suggestions that had been put forth about the structure of nucleic acids. In Levene's own words, "New facts and new evidence may cause its alteration, but there is no doubt as to the polynucleotide structure of the yeast nucleic acid" (1919).

Indeed, many new facts and much new evidence soon emerged and caused alterations to Levene's proposal. One key discovery during this period involved the way in which nucleotides are ordered. Levene proposed what he called a tetranucleotide structure, in which the nucleotides were always linked in the same order (i.e., G-C-T-A-G-C-T-A and so on). However, scientists eventually realized that Levene's proposed tetranucleotide structure was overly simplistic and that the order of nucleotides along a stretch of DNA (or RNA) is, in fact, highly variable. Despite this realization,

Levene's proposed polynucleotide structure was accurate in many regards. For example, we now know that DNA is in fact composed of a series of nucleotides and that each nucleotide has three components: a phosphate group; either a ribose (in the case of RNA) or a deoxyribose (in the case of DNA) sugar; and a single nitrogen-containing base. We also know that there are two basic categories of nitrogenous bases: the purines (adenine [A] and guanine[G]), each with two fused rings, and the pyrimidines (cytosine [C], thymine [T], and uracil [U]), each with a single ring. Furthermore, it is now widely accepted that RNA contains only A, G, C, and U (no T), whereas DNA contains only A, G, C, and T (no U) (Figure 1).

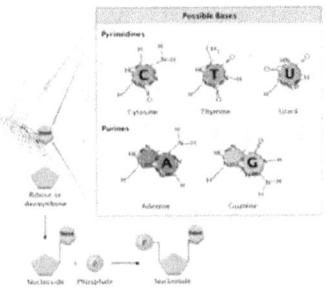

Figure 1: The chemical structure of a nucleotide.

A single nucleotide is made up of three components: a nitrogen-containing base, a five-carbon sugar, and a phosphate group. The nitrogenous base is either a purine or a pyrimidine. The five-carbon sugar is either a ribose (in RNA) or a deoxyribose (in DNA) molecule.

Strengthening the Foundation: Chargaff Formulates His "Rules"

Erwin Chargaff was one of a handful of scientists who expanded on Levene's work by uncovering additional details of the structure of DNA, thus further paving the way for Watson and Crick. Chargaff, an Austrian biochemist, had read the famous 1944 paper by Oswald Avery and his colleagues at Rockefeller University, which demonstrated that hereditary units, or genes, are composed of DNA. This paper had a profound impact on Chargaff, inspiring him to launch a research program that revolved around the chemistry of nucleic acids. Of Avery's work, Chargaff (1971) wrote the following:

"This discovery, almost abruptly, appeared to foreshadow a chemistry of heredity and, moreover, made probable the nucleic acid character of thegene... Avery gave us the first text of a new language, or rather he showed us where to look for it. I resolved to search for this text."

As his first step in this search, Chargaff set out to see whether there were any differences in DNA among different species. After developing a new paper chromatography method for separating and identifying small amounts of organic material, Chargaff reached two major conclusions (Chargaff, 1950). First, he noted that the nucleotide composition of DNA varies among species. In other words, the same nucleotides do not repeat in the same order, as proposed by Levene. Second, Chargaff concluded that almost all DNA--no matter what organism or tissue type it comes from--maintains certain properties, even as its composition varies. In particular, the amount of adenine

(A) is usually similar to the amount of thymine (T), and the amount of guanine (G) usually approximates the amount of cytosine (C). In other words, the total amount of purines (A + G) and the total amount of pyrimidines (C + T) are usually nearly equal. (This second major conclusion is now known as "Chargaff's rule.") Chargaff's research was vital to the later work of Watson and Crick, but Chargaff himself could not imagine the explanation of these relationships--specifically, that A bound to T and C bound to G within the molecular structure of DNA (Figure 2).

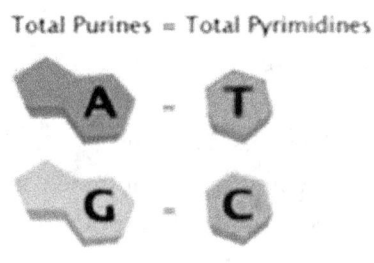

Figure 2: What is Chargaff's rule?

All DNA follows Chargaff's Rule, which states that the total number of purines in a DNA molecule is equal to the total number of pyrimidines.

Putting the Evidence Together: Watson and Crick Propose the Double Helix

Chargaff's realization that A = T and C = G, combined with some crucially important X-ray crystallography work by English researchers Rosalind Franklin and Maurice Wilkins, contributed to Watson and Crick's derivation of the three-dimensional, double-helical model for the structure of DNA. Watson and Crick's discovery was also made possible by recent advances in model building, or the assembly of possible three-dimensional structures based upon known molecular distances and bond angles, a technique advanced by American biochemist Linus Pauling. In fact, Watson and Crick were worried that they would be "scooped" by Pauling, who proposed a different model for the three-dimensional structure of DNA just months before they did. In the end, however, Pauling's prediction was incorrect.

Using cardboard cutouts representing the individual chemical components of the four bases and other nucleotide subunits, Watson and Crick shifted molecules around on their desktops, as though putting together a puzzle. They were misled for a while by an erroneous understanding of how the different elements in thymine and guanine (specifically, the carbon, nitrogen, hydrogen, and oxygen rings) were configured. Only upon the suggestion of American scientist Jerry Donohue did Watson decide to make new cardboard cutouts of the two bases, to see if perhaps a different atomic configurationwould make a difference. It did. Not only did the complementary bases now fit together perfectly (i.e., A with T and C with G), with each pair held together by hydrogen bonds, but the structure also reflected Chargaff's rule (Figure 3).

Discovery of DNA Structure and Function

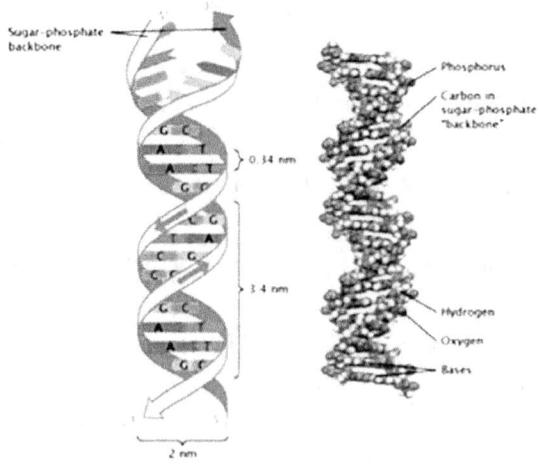

Figure 3: The double-helical structure of DNA.

The 3-dimensional double helix structure of DNA, correctly elucidated by James Watson and Francis Crick. Complementary bases are held together as a pair by hydrogen bonds.

Although scientists have made some minor changes to the Watson and Crick model, or have elaborated upon it, since its inception in 1953, the model's four major features remain the same yet today. These features are as follows:

• DNA is a double-stranded helix, with the two strands connected by hydrogen bonds. A bases are always paired with Ts, and Cs are always paired with Gs, which is consistent with and accounts for Chargaff's rule.

Discovery of DNA Structure and Function

- Most DNA double helices are right-handed; that is, if you were to hold your right hand out, with your thumb pointed up and your fingers curled around your thumb, your thumb would represent the axis of the helix and your fingers would represent the sugar-phosphate backbone. Only one type of DNA, called Z-DNA, is left-handed.

- The DNA double helix is anti-parallel, which means that the 5' end of one strand is paired with the 3' end of its complementary strand (and vice versa). As shown in Figure 4, nucleotides are linked to each other by their phosphate groups, which bind the 3' end of one sugar to the 5' end of the next sugar.

- Not only are the DNA base pairs connected via hydrogen bonding, but the outer edges of the nitrogen-containing bases are exposed and available for potential hydrogen bonding as well. These hydrogen bonds provide easy access to the DNA for other molecules, including the proteins that play vital roles in the replication and expression of DNA (Figure 4).

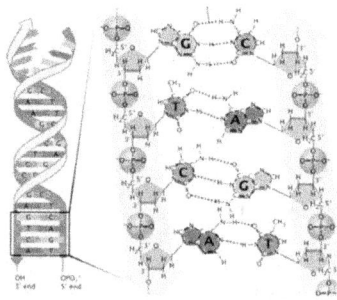

Figure 4: Base pairing in DNA.

Two hydrogen bonds connect T to A; three hydrogen bonds connect G to C. The sugar-phosphate backbones (grey) run anti-parallel to each other, so that the 3' and 5' ends of the two strands are aligned.

One of the ways that scientists have elaborated on Watson and Crick's model is through the identification of three different conformations of the DNA double helix. In other words, the precise geometries and dimensions of the double helix can vary. The most common conformation in most living cells (which is the one depicted in most diagrams of the double helix, and the one proposed by Watson and Crick) is known as B-DNA. There are also two other conformations: A-DNA, a shorter and wider form that has been found in dehydrated samples of DNA and rarely under normal physiological circumstances; and Z-DNA, a left-handed conformation. Z-DNA is a transient form of DNA, only occasionally existing in response to certain types of biological activity (Figure 5). Z-DNA was first discovered in 1979, but its existence was largely ignored until recently. Scientists have since discovered that certain proteins bind very strongly to Z-DNA, suggesting that Z-DNA plays an important biological role in protection against viral disease (Rich & Zhang, 2003

26

Discovery of Antibiotics

Alexander Fleming

Antibiotics - An Introduction

Infections are very common and responsible for a large number diseases adversely affecting human health. Most of the infectious diseases are caused by bacteria. Infections caused by bacteria can be prevented, managed and treated through anti-bacterial group of compounds known as antibiotics.

Definition

Antibiotics can be loosely defined as the variety of substances derived from bacterial sources (microorganisms) that control the growth of or kill other bacteria. However, Synthetic antibiotics, usually chemically related to natural antibiotics, have since been produced that accomplish comparable tasks.

Classifications

A common scheme of classifications for antibiotics is drawn below:

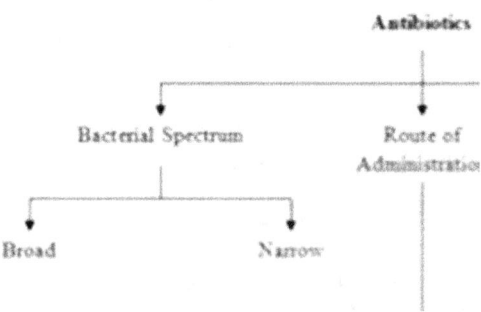

Antibiotics can also be classified based on their chemical structure. A similar level of effectiveness, toxicity and side-effects is rendered by the antibiotics of same structural group. Broad spectrum antibiotics are effective against a broad range of microorganisms in comparison to narrow spectrum antibiotics. Bactericidal antibiotics kill the bacteria whereas bacteriostatic antibiotics halt the growth of bacteria.

History of Antibiotics

History of antibiotics can be described in two segments as under:

Discovery of Antiobiotics

Early History

During ancient times;

• Greeks and Indians used moulds and other plants to treat infections.

• In Greece and Serbia, mouldy bread was traditionally used to treat wounds and infections.

• Warm soil was used in Russia by peasants to cure infected wounds.

• Sumerian doctors gave patients beer soup mixed with turtle shells and snake skins.

• Babylonian doctors healed the eyes using a mixture of frog bile and sour milk.

• Sri Lankan army used oil cake (sweetmeat) to server both as desiccant and antibacterial.

MODERN HISTORY

Year   Origin Description

1640   England   John Parkington recommended using mold for treatment in his book on pharmacology

Discovery of Antiobiotics

1870   England   Sir John Scott Burdon-Sanderson observed that culture fluid covered with mould did not produce bacteria

1871   England   Joseph Lister experimented with the antibacterial action on human tissue on what he called Penicilliumglaucium

1875   England   John Tyndall explained antibacterial action of the Penicillium fungus to the Royal Society

1877   France Louis Pasteur postulated that bacteria could kill other bacteria (anthrax bacilli)

1897   France Ernest Duchesne healed infected guinea pigs from typhoid using mould (Penicilliumglaucium)

1928   England   Sir Alexander Fleming discovered enzyme lysozyme and the antibiotic substance penicillin from the fungus Penicilliumnotatum

1932   Germany   Gerhard Domagk discovered Sulfonamidochrysoidine (Prontosil)

During 1940's and 50's streptomycin, chloramphenicol, and tetracycline were discovered and Selman Waksman used the term "antibiotics" to describe them (1942)

Discovery of Antiobiotics

Sir Alexander Fleming

Sir Alexander Fleming, a Scottish biologist, defined new horizons for modern antibiotics with his discoveries of enzyme lysozyme (1921) and the antibiotic substance penicillin (1928). The discovery of penicillin from the fungus Penicilliumnotatum perfected the treatment of bacterial infections such as, syphilis, gangrene and tuberculosis. He also contributed immensely towards medical sciences with his writings on the subjects of bacteriology, immunology and chemotherapy.

Alexander Fleming was born in Loudon, Scotland on 6 August, 1881 in a farming family. He carried on his schooling at Regent Street Polytechnic after his family moved to London in 1895. He joined St. Mary's Medical School and became research assistant to renowned Sir Almroth Wright after he qualified with distinction in 1906. He completed his degree (M.B.B.S.) with gold medal in 1908 from the University of London and lectured at St. Mart till 1914. He served as Captain during the World War I and worked in battlefield hospitals in France. After the war he returned to St. Mary in 1918 and got elected Professor of Bacteriology in 1928.

The Discovery of Antibiotics

"One sometimes finds what one is not looking for"

(Sir Alexander Fleming)

His research and study during his military career inspired him to discover naturally antiseptic enzyme in 1921, which he named lysozyme. This substance existed in tissues and secretions like mucus, tears and egg-white but it did not have much effect on the strongly harmful bacteria. Six years later; as a result of some intelligent serendipity, he stumbled on discovering penicillin. It was in 1928 when he observed while experimenting on influenza virus that a common fungus, Penicilliumnotatum had destroyed bacteria in a staphylococcus culture plate. Upon subsequent investigation, he found out that mould juice had developed a bacteria-free zone which inhibited the growth of staphylococci. This newly discovered active substance was effective even when diluted up to 800 times. He named it penicillin.

He was knighted in 1944 and was given the Nobel Prize in Physiology or Medicine in 1945 for his extraordinary achievements which revolutionized the medical sciences.

How Do Antibiotics Work?

Various types of antibiotics work in either of the following two ways:

1. A Bactericidal antibiotic kills the bacteria generally by either interfering with the formation of the bacterium's cell wall or its cell contents.

Penicillin, daptomycin, fluoroquinolones, metronidazole, nitrofurantoin and co-trimoxazole are some example of Bactericidal antibiotics.

2. A Bacteriostatic antibiotic stops bacteria from multiplying by interfering with bacterial protein production, DNA replication, or other aspects of bacterial cellular metabolism.

Some Bacteriostatic antibiotics are tetracyclines, sulphonamides, spectinomycin, trimethoprim, chloramphenicol, macrolides and lincosamides.

https://explorable.com/history-of-antibiotics

27

THE FOUNDING OF GENETICS

Gregor Mendel

Edited notes by Jill Hunter (1996) and Rita Mitchell (1997

Genetics and Inheritance

Gregor Mendel's research with genes and the way they work was the main starting point towards future research in the field of genetics. Mendel first entered a monastery in order to "free [himself] from the bitter struggle for existence." The Abbot of this monastery had an interest in plant breeding- particularly pea plants - and organized a course in crossbreeding during Mendel's first year at the monastery. Mendel was intrigued by the results of crossbreeding. After his participation in the Abbot's course, he left the monastery to study math and physics at the University of Vienna. Two years later he returned to the monastery, where he began his own experiments with pea plants.

The Founding of Genetics

In 1865, Mendel published the paper "Experiments in Plant Hybridization", which appeared in the journal of a local natural history society. This paper showed that each organism has physical traits that correspond to invisible elements within the cell. These invisible elements, which we now call genes, exist in pairs. Mendel showed that only one member of this genetic pair is passed on to each progeny. The gene is incorporated into a sperm or egg cell. At the time that Mendel did his research, there was no knowledge of chromosomes, cell structure, fertilization, mitosis and meiosis, which is now taught as common knowledge to most school children. Mendel's findings, therefore, show a remarkable use of observation and deduction which was quite ahead of his time. In 1868, the Abbot died. Mendel then ran the monastery, thus stopping all of his scientific work. However, he had laid the groundwork for the field of genetics.

Geneticists distinguish between genes and their expression, as it is then easier to discuss experimental results. A genotype is the actual set of genes that an organism has; it is the blueprint of genetic material. A phenotype is a measurable characteristic of an organism, such as eye color, hair color, the shape of one's nose, number of fingers, behavioral traits such as smiles per hour, physiological traits such as heart rate, or biochemical traits such as cholesterol levels and blood type.

In Mendel's studies, he used peas because they:

- are easy to culture

- produce many offspring per mating

- are capable of self-fertilization

- have a number of varieties in type, as well as in phenotype.

Mendel was extremely careful in gathering his data and scrutinizing the results of every experiment. This attention to detail, along with his mathematical insight, allowed Mendel to keep meticulous data.

When doing his work, Mendel posed questions dealing with how the physical characteristics of peas were transmitted from generation to generation, and whether these transmissions were unchanged or altered when passed on. He also questioned whether hereditary particles existed. His studies were based on seven traits of peas. The seven traits were qualitative (they could be measured and a value assigned); therefore, specified qualities could be assigned to each plant. These characteristics were visible and it was through them that he could study the effects of reproduction.

The seven traits were:

1. shape of the seed - round or wrinkled.

2. color of the pea - yellow or green.

3. color of the seed - gray or white.

4. form of the ripe pod - inflated or constricted between peas.

5. color of the unripe pod - green or yellow.

6. position of the flower - terminal or axial.

7. length of the stem - tall or short.

Mendel began his experiments using a set of pure-breeding pea plants, meaning that the second generation of plants had consistent traits with those of the first. He performed monohybrid crosses, meaning that the experiment was carried out between two strains of plants that differed only in one characteristic. He crossed parents of different phenotypes to see what resulted. The parents were denoted by a P, while the offspring - thefilial generation - was denoted by F1, the next generation F2, etc.

Mendel proceeded to conduct a series of monohybrid crosses (testing one trait at a time), meaning that for each of the seven phenotypes, he crossed two plants of opposite phenotypes. Consistently, Mendel found that in the first generation of these crosses, all of the F1s were identical to one of the parents. For example, when testing the shape of the seed, crossing one pure-bred round seed with a pure-bred wrinkled seed, all of the offspring were round. The one trait - in this case a wrinkled seed - not expressed in the offspring he called a recessive trait. In each case of this crosses, the round trait was dominant over the wrinkled trait and is said to be the dominant trait. This conclusion is now referred to as Mendel's Law of Dominance. During the experiments, Mendel also observed that the sex of the parent was irrelevant for the dominant or recessive trait exhibited in the offspring. A cross between a male round seed with a female wrinkled seed would offer identical results to a female round seed crossed with a male wrinkled seed. This is known as Mendel's Law of Parental Equivalence.

Mendel found new principles in that the phenotypes absent in the F1 generation reappeared in approximately 1/4 of the F2 offspring. Mendel could not predict what traits would be present in any one individual, but he did deduce that there was a 3:1 ratio in the F2 generation for dominant/recessive phenotypes.

In describing his results, Mendel used the term elementen, which he postulated to be hereditary particles transmitted unchanged between generations. Even if the traits are not expressed, he surmised that they are still held intact. For example, even if a trait is not expressed (such as the wrinkled seed in the first filial generation of Mendel's crossing), that plant still has a wrinkled seed allele; the elementen are still passed on. We now call these 'particles' alleles. An allele that can be suppressed during a generation is called a recessive allele, while one that is consistently expressed is a dominant allele.

We have developed terms to describe the existence of recessive and dominant alleles in any given genotype. Homozygous plants or individuals are those who have two copies of the same gene, e.g. Round/Round or Wrinkled/Wrinkled. Heterozygous individuals received a different type of gene from each zygote, eg. Round/Wrinkled or Wrinkled/Round. In this case, Round being the dominant trait, the phenotype of the plant would be Round, though the genotype Round/Wrinkled.

To help clarify the difference between a gene and an allele, think of it this way. A gene is a location on a chromosome. Alleles are different options for the same gene. For example, there may be a specific gene for eye color - meaning a location on the chromosome at which eye color is specified. Whichever allele (for green eyes, blue eyes, brown eyes) gets placed in that location will determine the specific color of the eyes.

Mendel observed that although the dominant trait was the one expressed in the F1 generation, the recessive trait still had an effect on the genotype of a heterozygote's offspring. Mendel developed a kind of shorthand for distinguishing alleles. They are designated by one or more letters. The first letter of abbreviation of a dominant allele is uppercase, for instance Rfor round seeds, and the first letter of abbreviation for a recessive allele is lowercase, i.e. rfor wrinkled seeds. Thus, two alleles can result in three different genotypes: RR, Rr, and rr. RR and Rr have the same phenotype, however, because the dominant trait is the one expressed.

Sometimes mating results can be expressed in the form of a Punnett Square.

	R	r
R	RR	Rr
r	rR	rr

Each box in the top row represents the gametes from one of the parents, while the left column shows the gametes from the other parent.

There is a 25% chance for each type of crossing (if there are only two gametes). However, though there are four different combinations, there are only two phenotypes (dominant or recessive), thus explaining Mendel's results. The ratio of dominant to recessive phenotypes is 3:1 - Mendel's results! Mendel's data was significant and compelling because he expressed it in ratios that always held true.

Mendel's hypothesis also allowed a prediction to be made that could be tested. He postulated that there were three different genotypes in F2 (RR, Rr, and rr). Individuals with recessive (wrinkled) phenotype would produce only wrinkled individuals when self-fertilized in the F3 generation, since recessive homozygotes would breed true. He also assumed that round individuals would be either RR or Rr, in the ratio of 1:2. If these individuals were self-fertilized, then:

• 1/3 (RR) should breed true and produce only round seeded plants

• 2/3 (Rr or rR) should produce both round and wrinkled peas in the ratio of 3:1

This model explained the results of past experiments, but also rendered a means to predict future results.

Mendel's Law of Segregation states that each member of a pair of alleles maintains its own integrity, regardless of which one(s) is dominant. At reproduction, only one allele of a pair is transmitted to each gamete, and that choice is entirely random.

After his monohybrid crosses, Mendel did a series of dihybrid crosses. These were crosses between strains identical except for two characteristics. For example, Mendel did experiments with round/wrinkle (R/r) and yellow/green (Y/y). Mendel crossed a double dominant P (RRYY) with a double recessive P (rryy). All of the F1s that resulted from this were thus RrYy and had the characteristics of round and yellow. These double heterozygotes were then used to make an F2 generation. The examination of one trait at a time demonstrates a 3:1 ratio:

- round:wrinkle (423:133 = 3.18:1)

- yellow:green (416:140 = 2.97:1)

Mendel thus observed that each of the traits he was following sorted themselves independently. Mendel's Law of Independent Assortment states that characteristics which are controlled by different genes will assort independent of all others. Whether or not a seed will be Rr or RR has nothing to do with whether or not it will be Yy or yy.

Some experiments produce data different from that predicted by Mendel's Laws

Although Mendel did not do experiments with humans, we know now that humans have approximately 100,000 genes.

The Founding of Genetics

The reason that we are all so different and can even look so different from both of our biological parents is because there are so many different genetic combinations that can result. It is possible to generate about 8 million different types of gametes from a person who is heterozygous for only one gene on each pair of 23 chromosomes!

These laws Mendel discovered for the most part hold true. However, there are instances where Mendelian rules become complicated. Sometimes, within a population, more than two alleles can occur at a particular locus. This is multiple allelism and occurs in such instances as the ABO blood group which has at least three common alleles and some that are more rare. The HLA-A antigen group has at least 23 different alleles.

In some heterozygous organisms, both alleles of a given locus are expressed, showing codominance. For example, the ABO blood group shows codominance. If a red blood cell as an "A" allele it will express the "A" protein on its surface. If it has a "B" allele it will express the "B" protein on its surface. If both alleles are present, then both surface proteins will be present. In this case, neither allele is dominant over the other.

When one characteristic is reduced in a heterozygous organism, the alleles are said to exhibit partial or incomplete dominance because two copies of the gene will produce more or an effect than only a single copy of the gene. We see partial dominance in nature is some flower colors. When we see pink flowers, the allele for red is present only one time instead of two, resulting in the presence of only half the amount of red pigment.

Not all inheritance is not dependent solely on the copy of a single gene. Some characteristics, such as height, depend on the way in which many of the genes interact in an organism. It is, then, a polygenic trait.

Linkage

The Law of Independent Assortment described above does not apply to all situations. On a given chromosome, there are sets of genes, and genes that are on the same chromosome are physically linked to one another. During reproduction, these linked genes tend to be transmitted as a unit instead of independently - this is referred to as linkage and these genes are said to be linked. Although these links may break during meiosis, this is not always the case. Therefore, the Law of Independent Assortment does not always hold true. Independent assortment sometimes occurs when genes are located on the chromosome but are relatively far apart from one another making the occurrence of crossover between the two locations likely. Similarly, mutations sometimes occur in a gene. A mutation simply is a change in the DNA sequence of a gene. It can be due to a change in a base (letter) of the gene or due to an insertion or deletion of a piece of DNA. Although mutations occur in the normal course of DNA replication, the probability of one gene changing in one cell cycle is less than one in a million in humans. Because these occurrences are so rare, Mendel did not have any problems with his results.

https://www.dartmouth.edu/~cbbc/courses/bio4/bio4-1997/01-Genetics.html

28

The Microprocessor

Ted Hoff

By Iain Mackenzie

Ted Hoff saved his own life, sort of.

Deep inside this 73-year-old lies a microprocessor - a tiny computer that controls his pacemaker and, in turn, his heart.

The Microprocessor

Microprocessors were invented by - Ted Hoff, along with a handful of visionary colleagues working at a young Silicon Valley start-up called Intel.

This curious quirk of fate is not lost on Ted."It's a nice feeling," he says.

Memory

In 1967 Marcian Edward Hoff decided to walk away from academia, having gained his PhD in electrical engineering.

Robert Noyce, the "Mayor of Silicon Valley", head-hunted Ted Hoff for his new company

A summer job developing railway signalling systems had given him a taste for working in the real world. Then came a phone call that would change his life.

The Microprocessor

"I had met the fella once before. His name was Bob Noyce. He told me he was staffing a company and asked if I would consider a position there," says Ted.

Six years earlier, Robert Noyce, the founder of Fairchild Semiconductor, had patented the silicon chip. Now his ambitions had moved on and he was bringing together a team to help realise them.

"I interviewed at Bob Noyce's home and he did not tell me what the new company was about," says Ted."But he asked me if I had an idea what the next level for integrated circuits would be and I said, 'Memory' ".

He had guessed correctly. Mr Noyce's plan was to make memory chips for large mainframe computers. Ted was recruited and became Intel employee number 12.

In 1969, the company was approached by Busicom, a Japanese electronics maker, shopping around for new chips. It wanted something to power a new range of calculators and asked for a set-up that used 12 separate integrated circuits.

Ted believed he could improve on that by squashing most of their functions onto a single central processing unit. The result was a four-chip system, based around the Intel 4004 microprocessor.

Sceptics

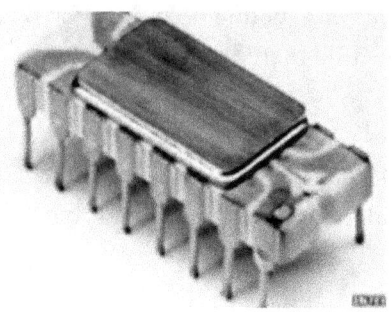

The four-bit 4004 microprocessor was originally designed for a Busicom calculator

Intel's work was met with some initial scepticism, says Ted. Conventional thinking favoured the use of many simple integrated circuits on separate chips. These could be mass produced and arranged in different configurations by computer-makers. The entire system offered economies of scale.

But microprocessors were seen as highly specialised - designed at great expense only to be used by a few manufacturers in a handful of machines. Time would prove the sceptics to be 100% wrong.

Intel also faced another problem. Even if mass production made microprocessors cheaper than their multiple-chip rivals, they were still were not as powerful. Perhaps early computer buyers would have compromised on performance to save money, but it was not the processors that were costing them.

The Microprocessor

"Memory was still expensive," says Ted."One page of typewritten text may be 3,000 characters. That was like $300 [£182].

"If you are going put a few thousand dollars worth of memory [in a computer], wouldn't it make more sense to spend $500 for a processor built out of small- or medium- scale electronics and have 100 times the performance.

"At that time, it didn't really make sense to talk about personal computers," he said.

Over time, the price of computer memory would began to fall and storage capacity increase.Intel's products started to look more and more attractive, although it would take another three years and four chip generations before one of their processors made it into a commercially available PC.

Moore's law

Intel knew its system would win out eventually.It could even predict when microprocessors would make the price-performance breakthrough.

In 1965, Gordon Moore, who would later co-found Intel with Robert Noyce, made a bold prediction.

Gordon Moore's law continues to hold, almost half a century after he proposed it

He said: "The complexity for minimum component costs has increased at a rate of roughly a factor of two per year".

The theory, which would eventually come to be known as Moore's Law, was later revised and refined. Today it states, broadly, that the number of transistors on an integrated circuit will double roughly every two years. However, even Mr Moore did not believe that it was set in stone forever.

"Gordon always presented it as an observation more than a law," says Ted. Even in the early days, he says, Intel's progress was out-performing Moore's law.

Ubiquitous chips

As the years passed, the personal computer revolution took hold. Microprocessors are now ubiquitous. But Ted believes the breadth of their versatility is still under-appreciated.

"One of the things I fault the media for is when you talk about microprocessors, you think about notebook and desktop computers.

The Microprocessor

"You don't think of automobiles, or digital cameras or cell phones that make use of computation," he says.

Ted launches into an awed analysis of the processing power of digital cameras, and how much computing horsepower they now feature.Like a true technologist, the things that interest him most lie at the bleeding edge of electronic engineering.

Attempts to make him elevate his personal achievements or evaluate his place in history are simply laughed off.

Ted Hoff (far left) receives the National Medal of Technology and Innovation in 2010

Instead, Ted would rather talk about his present-day projects.

"I have a whole bunch of computers here at home. I still like to play around with micro-controllers.

"I like to programme and make them solve technical problems for me," he says.

But if Ted refuses to recognise his own status, others are keen to.

In 1980 he was named the first Intel Fellow - a position reserved for only the most esteemed engineers.Perhaps his greatest honour came in 2010 when US President Barack Obama presented Ted with the National Medal of Technology and Innovation.

His name now stands alongside other winners including Gordon Moore, Robert Noyce, Steve Jobs, Bill Gates and Ray Dolby.Like them, he helped shape the world we live in today.

http://www.bbc.co.uk/news/technology-13260039

29

Quantum Theory

Max Planck

Born in 1858, Max Planck came from an academic family. His father Julius Wilhelm Planck was Professor of Law at the University of Kiel, Germany, and both his grandfather and great-grandfather had been professors of theology at Göttingen.

Quantum Theory

He began his elementary schooling in Kiel but in 1867 his family moved to Munich, where his father was appointed Professor. The city provided a stimulating environment for the young boy who enjoyed its culture, particularly the music, and loved walking and climbing in the mountains when the family took excursions to Upper Bavaria.

He was not outstanding at school but did well enough to enter the University of Munich on 21 October 1874, where started taking mathematics classes and then decided to study physics.

It was customary for German students to move between universities at this time and Planck studied at the University of Berlin from October 1877, where his teachers included Helmholtz and Kirchhoff. He returned to Munich and received his doctorate in July 1879 at the age of 21 submitting a thesis on the second law of thermodynamics.

In 1885, he was appointed 'Extraordinary Professor of Theoretical Physics' in Kiel. After the death of Kirchhoff in October 1887, the University of Berlin looked for a physicist to replace him. Planck was proposed by Berlin's Faculty of Philosophy and strongly recommended by Helmholtz. He was promoted to Ordinary Professor in 1892 and held the chair until he retired in 1927. He continued to indulge his passion for music, having a special harmonium built and holding concerts in his own home.

While in Berlin, he became fascinated by the way energy from hot objects was emitted in variable quantities depending upon wavelength. A number of physicists had tried to find a mathematical description but no one had been totally successful. By combining equations derived by Wien and Rayleigh, in October 1900 Planck announced a result now known as Planck's radiation formula.

Within two months, he explained why his formula worked – and it was a bold explanation. He renounced previous physics and introduced the concept of 'quanta' of energy. These are small 'packets' that can only hold certain, prescribed amounts of energy.

In December 1900, he presented his theoretical explanation involving quanta at a meeting of the PhysikalischeGesellschaft in Berlin. In doing so, he had to reject his belief that the Second Law of Thermodynamics was an absolute law of nature.

At first his theory met resistance but, due to the successful work by Niels Bohr in 1913 calculating positions of spectral lines using the theory, it became generally accepted. The quantum theory was born. Planck himself said that, despite having invented quantum theory, he did not understand it himself at first. Nevertheless, he received the Nobel Prize for Physics in 1918 for his achievement.

Planck was 42 years old when he made his historic quantum announcement, but took only a minor part in the further development of quantum theory. This was left to Einstein, Poincaré, Bohr, Dirac and others.

His private life was filled with tragedy. His first wife died in 1909, his eldest son was killed in 1916 during World War I, and both his daughters died in childbirth. His home in Berlin was destroyed by fire after an air raid in February 1944, and his second son was suspected of involvement in the plot to assassinate Hitler and executed in 1945.

Planck was 87 years old by the end of World War II but, remarkably, he was able to put effort into reconstructing German science as president of the Kaiser Wilhelm Gesellschaft. He died in October 1947.

Named in his honour, ESA's Planck mission is now analysing the Cosmic Microwave Background radiation, which was formed soon after the Big Bang. The data Planck collects will allow astronomers to search for clues about how galaxies form and cluster together, giving us the large-scale structure we see around us in space today.

http://www.esa.int/Our_Activities/Space_Science/Planck/Max_Planck_Originator_of_quantum_theory

30

THE INVENTION OF THE PRINTING PRESS

Johann Gutenberg

Movable Type

The printing press invented by German goldsmith Johann Gutenberg in 1448 has been called one of the most important inventions in the history of humankind. The device made it possible for the first time for the common man, woman and

The Invention of The Printing Press

child to have access to books, which meant they would for the first time have unprecedented ability to accumulate knowledge.

Before the invention of the printing press, the majority of books were written and copied by hand. Block printing was becoming more popular, which involved carving each page of a text into a block of wood and pressing each block onto paper. Because these processes were so labor-intensive, books were very expensive, and only the rich could afford them.

Believed to have been born in Mainz, Germany, in approximately 1399, Gutenberg, nee Johann Gensfleisch, later adopted his family's settling place as his last name. He was trained as a goldsmith, gem cutter and metallurgist. For some time he lived in Strasbourg, most likely in the late 1430s-early 1440s. By then he had been losing money in his business and began looking for a way to make money to pay off his debts.

He started working on a device that would make it possible to print texts using movable blocks of letters and graphics. These, used with paper, ink, and a press, would make it possible to print books much faster and more cheaply than ever before. He used metals he was familiar with ¬ lead, antimony and tin, to cast 290 blocks of letters and symbols and created a linseed- and soot-based ink of the consistency he believed to be ideal for printing on handmade paper. He adapted a wine press that allowed him to slide paper in and out of it and to squeeze water from the paper after printing.

He tested his moveable-type machine by printing a Latin book on speech-making in 1450. When this endeavor was successful,

The Invention of The Printing Press

he embarked on his most famous project, the printing of The Gutenberg Bibles.

The bibles, printed in Latin, gained fame as the first books ever printed in Europe, and the first bibles printed in history. Two-hundred copies were made, each complete with beautiful illustrations and vibrant colors. Part of Gutenberg's genius was his technique for creating blocks to represent the calligraphy done in hand-made volumes, so that the richness of the original texts could be preserved. Characters and illustrations were later hand-illuminated. Today, only 22 of the original Gutenberg bibles are known to be in existence.

Gutenberg's business partner Johann Fust eventually gained ownership of the printing business and completed the printing of the bibles. This was the result of a deal made between the two men necessitated by debts Gutenberg owed to Fust. Gutenberg died in approximately 1468 in Mainz.

It should be noted that others in history claim to have come up with the idea of movable type earlier than Gutenberg did, including a Dutchman and a Chinese inventor. A system similar to his to said to have also been used in the 12th century in Korea. But for whatever reason Gutenberg's endeavor was the first to be successful and indeed, his printing press had a revolutionary impact on history and the entire world.

The printing press and all that it brought to the masses helped to inspire a religious revolution as families were for the first time able to possess a Bible for their own interpretation. It also factored into the progress of science, general education, and is said to have been key in moving the world out of the Medieval era into the Early Modern period.

31

A SHORT HISTORY OF THE INTERNAL COMBUSTION ENGINE

Valerie Strauss

Wheels keep on turnin'

How many others have succeeded, but failed? It seems that history has more holes than a wheel of Swiss cheese. I discovered the Petrol-Cycle on the front page of the February 14, 1891 edition of the Scientific American. Invented by Edward Butler of Greenwich, England, this three-wheeled fore-car had an elegant 650cc twin-cylinder, four-stroke, water-cooled engine with electric spark ignition. Years ahead of DeDion and Boulton, the Petrol-Cycle made Gottleib Daimler's Reitwagen look like something from the medieval period. Regardless, pick up any reference book on the history of motorcycles and you won't find it mentioned.

A Short Hisroty of The Intenral Combustion Engine

Built in 1888 by the Merryweather Fire Engine Company in Greenwich, England, the Petrol-Cycle was driven by a 5/8 hp engine with a magneto ignition (later replaced by a coil and battery) and could attain a speed of 10 mph. It had Ackermann steering, rotary valves, and a float-fed carburetor—five years before Wilhelm Maybach invented his spray carburetor in 1893. The liquid-cooled engine had a water reservoir that doubled as a rear fender (an idea adopted by Hilderbrand&Wolfmuller in 1894) and the engine was started using compressed air.

Being ahead of your time isn't always a good thing. The problem with Bulter's creation wasn't mechanical, but political. The English Red Flag Act (Locomotives On Highways Act) of 1865 limited all self-propelled vehicles to a speed limit of 2 mph in towns and 4 mph in rural areas. In addition, it required three people to attend to the vehicle, one of which had to walk in advance of the vehicle waving a red flag to warn people of its approach. In the 1890 issue of The English Mechanic, he wrote, "The authorities do not countenance its use on the roads, and I have abandoned in consequence any further development of it." In 1896 he broke the Petrol-Cycle up for scrap and sold his patents to Harry Lawson, who manufactured the engine for marine use. Ironically, the Red Flag Act was repealed that very year.

Other than the steam velocipedes of Sylvester Roper, the history of motorcycles doesn't begin until after the internal combustion engine had evolved. Count Albert DeDion and George Boulton had been producing steam vehicles since 1883, but when they developed their experimental internal-combustion engines, everything changed. The DeDion 1/2 hp,

A Short Hisroty of The Intenral Combustion Engine

four-stroke, 138cc engine powered their first line of tricycles in 1895 and it was the franchising of this engine that gave birth to the modern motorcycle.

Sylvester Roper was a noted inventor who had the finances, access to state-of-the-art fabricating facilities, and extensive marketing connections to develop his passion for motorcycling. At least two of his business associates built and rode steam wagons as did Roper prior to 1867, and he actively promoted his steam velocipedes right up to the time of his death. However, a number of diverse elements had to be developed before motorcycles could become viable. One of these was the internal combustion engine.

History acknowledges that Gottlieb Daimler and Wilhelm Maybach built the first motorcycle with an internal combustion (IC) engine in 1885, and that Karl Benz built his two-stroke engine in 1879 to propel his three-wheeled Motowagen. Both Diamler and Maybach had worked with Nikolaus Otto, who developed the first successful gas engine. In fact, N.A. Otto &Cie is still in business. The story of the internal combustion engine didn't begin in 1876 when Nikolaus Otto developed his four-stroke gas engine and Sir Dougald Clerk built the first successful two-stroke engine. Siegfried Marcus built four operational cars—the Strassenwagen—between 1868 and 1873, but it was Étienne Lenoir who developed the first practical gas engine in 1860. Running on benzine and utilizing electric-spark ignition, Lenoir's engine powered a vehicle that he drove from Paris to Joinville in 1862.

A Short Hisroty of The Intenral Combustion Engine

Charles and Frank Duryea are recognized for manufacturing the first gasoline-powered cars in America (1893). Charles credits Samuel Morey for inventing the internal combustion engine and in the 1930s he financed the construction of two engines based on Morey's patent. One is in the Smithsonian and Dean Kamen owns the other.

Like Edward Butler, Captain Samuel Morey was a man ahead of his time. He built and successfully operated steamboats years before Robert Fulton, and was the first to use side paddles for propulsion. He patented a rotary steam engine, developed a new method of producing "lamp gas," patented a successful hot water heater using the same principle as modern ones, and then turned his attention to combustible gases. He not only built a "vapor engine," but also installed it in a boat and a wagon in 1829. Little is known about the successful boat demonstration on the Schuylkill River in September, but he made history on Market Street in Philadelphia by being the first person in the United States to operate a gas-powered vehicle, and the second in the world to have an automotive accident.

Unfortunately, Samuel Morey's patent was lost in the great Patent House Fire of 1836 and while Charles Duryea obviously had access to these papers, they once again became "lost" until being rediscovered in the archives of Dartmouth College in 2004. Utilizing valves, cams, a carburetor, and two cylinders, it is remarkably similar to internal combustion engines that would appear a half-century later. It ran on highly flammable spirits of turpentine and even had a wire mesh to prevent sparks in the combustion chamber from reaching the carburetor, a detail that would be independently reinvented in 1872. Samuel Morey was widely recognized for his accomplishments during his lifetime, yet he was unable to sell one of his most advanced patents and it became lost to history not once, but twice!

Samuel Morey's engine predates the first commercial steam locomotive in America, but he wasn't the first to experiment with internal combustion. A few months before Morey began designing his vaporengine, Samuel Brown patented an internal combustion engine on December 4, 1823 and built two working examples that were demonstrated at his home in Eagle Lodge in Brompton, England. He designed another engine in 1826 that used hydrogen and oxygen. Similar to a Newcomen steam engine, it had separate combustion and working cylinders, but water was circulated around the cylinder using a pump, and this water was cooled by contact with the outside air much like a modern water-cooled engine. This massive engine had a displacement of 8,800cc. Another source states the three cylinders had a bore of 12 inches and a stroke of 24 inches, but delivered only 4 hp. He mounted this on a four-wheeled vehicle and, in front of a small crowd, drove it up Shooter's Hill in Greenwich on the morning of May 27, 1826.

Samuel Morey's engine was much smaller and far more sophisticated than Brown's, but both were, like Newcomen's atmospheric steam engine, designed to create vacuum as the driving force. Still, neither of these men were the first to build operating IC engines, nor were they the first to drive a vehicle powered by such.

Based on a paper read before the Cambridge Philosophical Society in 1820, it's apparent that Rev. W. Cecil built an experimental IC engine. The title of this paper, "On the Application of Hydrogen Gas to Produce A Moving Power In Machinery, With A Description of an Engine Which Is Moved By the Pressure of The Atmosphere Upon A Vacuum Caused By Explosions of Hydrogen Gas and Atmospheric Air," seems to say it all. Rev. Cecil claimed that his engine ran at a steady 60 rpm and consumed 17.6 cu/ft of hydrogen per hour. The explosion took place in a cylinder and the engine apparently was quite noisy as he states: "To remedy the noise which is occasioned by the explosion, the lower end of the cylinder A, B, C, D may be buried in a well or it may be enclosed in a large air-tight vessel." He also mentioned an engine that Professor Farish, a lecturer on mechanics at Cambridge, had demonstrated. The Farish engine ran on gas and air that mixed in a cylinder and exploded by atmospheric pressure, but he apparently built an earlier engine that was fired by gunpowder. These are the earliest internal combustion engines—or were they?

Looking through titles to U.S. patents filed prior to the 1836 fire at the patent office I notice several atmospheric engines listed. Unfortunately there's no way to determine whether these were steam or gas engines unless they include the word "explosive." There's also no way to determine whether or not working engines were built for these patents.

One such example in 1811 was granted to Augustus Day of Bordentown, New Jersey, for his Combustion & Explosion Motor. He made improvements and filed another patent titled Explosive Power Motor on April 8, 1812. There are reputed to be published reports of this engine having actually been built. Of course, not all inventions were patented, nor all endeavors published. It has been alleged that John Sullivan had Marc Brunel build a gas-powered engine that was used to propel a boat in Hoboken, New Jersey, in 1798 or 1799. Considering the achievements of both men, this claim is plausible and might be one of those forgotten events in the history of motorized transportation.

Experiments with internal combustion were taking place in Europe even as James Watt's new steam engine was starting to be commercialized. John Barber's patent of 1791 was for an engine powered by the combustion of hydrocarbons and air. The hydrocarbons would be created by the external burning of wood, coal, oil, or other combustibles and then mixed with air in a second container he called the "exploder." The flash from the ignited gas would turn a paddlewheel or fan. Robert Street designed a compressionless piston engine in 1794. His proposal had coal tar, spirit or turpentine rendered gaseous and then ignited by a flame burning outside the cylinder. The combustion forced the piston upwards and moved a weight. Philippe Lebond'Humbersin patented, but never built, a machine driven by illuminating (coal) gas in 1801. Two pumps compressed air and gas separately, and these were released into a vessel where the mixture was to be ignited. The expansion of these gases overflowed into a double-acting cylinder and provided working power on both ends of the cylinder. He even

A Short Hisroty of The Intenral Combustion Engine

suggested using electricity for the ignition, and that both the generation of this electricity and the two compression pumps be run by the motor itself.

Francois Issac de Rivaz, a Swiss inventor, patented an internal-combustion engine in 1807. The engine was fueled by hydrogen and oxygen, stored in a balloon that was electrically ignited by a hand-operated trigger. Fitted to a crude four-wheeled wagon, the first test drive of 100 meters took place in 1813, thus making history as the first vehicle known to have been powered by an internal-combustion engine.

Like layers of an onion, the history of internal combustion keeps getting peeled back layer by layer. William MacGregor, in his 1885 work on gas engines, gives credit for their invention to the famous inventor and philosopher Abbe Hautefeuille in 1678. Most historians agree that in 1673 ChristiaanHuyghens was the first to describe a cylinder and piston in a "powder" machine, but there's no record of this machine having been more than a concept on paper. On January 16, 1206, the famous artist/ scholar/ astronomer/mechanical engineer al-Jaziri published his Book of Knowledge of Ingenious Devices. Among the 50 mechanical devices illustrated, he presents an ingenious water pump. The diagram looks like a 1950s vintage straight-six engine and is actually the earliest known documentation of a camshaft, reciprocal pistons, and sequenced valve system. While al-Jaziri's water pump had nothing to do with internal combustion, it was the form that function would ultimately follow seven centuries later.

http://www.thunderpress.net/motorcycle-touring/a-short-history-of-the-internal-combustion-engine/2009/01/19.htm

32

The Invention of Paper

Tsai Lun

A lot of the people living in the modern age take paper for granted. It is a common object found nearly everywhere, it is cheap and light, and it is easy to get a hold of. But paper has come a long way from its creation to the form we know today. People may not know it but the invention of paper has revolutionized a lot of things like the very civilization and culture and education of people.

The Invention of Paper

According to recent research and excavations, the earliest form of paper was dated back to the Western Han Dynasty, but this type of paper was made hemp that was pounded and disintegrated. It was very coarse, had an uneven texture and it was very thick. This type of paper was unearthed in a Han tomb somewhere in Ganshu Province and it is so far the earliest type of paper found.

But during the Eastern Han Dynasty around 104 A.D., a eunuch of the Imperial Court named CaiLun invented a new type of paper. It was said that he took bamboo fibers and the inner bark of a mulberry tree. He then added water to these and pounded them using a wooden tool. When they were pounded thoroughly, he poured the whole mixture over a flat woven cloth letting the water drain out. When it was dried, only the fibers remained and with this, CaiLun realized he had made a material that has a good writing surface and that it was lightweight. It was also very easy to make. CaiLun used other materials for his paper making, such as remnants or hemp, tree barks, fishnets and linen rags. In 105 A.D., he presented this invention to He Di, the emperor at that time and paper was then invented, according to Chinese History.

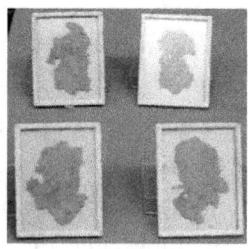

Photo by: Wikipedia Creative Commons

The Invention of Paper

Before CaiLun invented paper, writing surfaces were made from different materials such as bones, bamboo slips, wooden boards and even tortoise shells. These things are not only heavy but they also took up a lot of space and are hard to carry around. People then needed not only intelligence to study, but they needed to be strong to carry their books as well. Because of this, many thought these kinds of writing surface were unsuitable. It was probably what prompted CaiLun to invent a new lightweight writing surface that wasn't too thick or too bulky.

This new art of paper making later spread to East Korea about 384 A.D. Then around 610 A.D., a Korean monk took his paper making skills with him to Japan. And during the Tang Dynasty war, the Arab Empire captured soldiers and also some paper making workers so the skills were brought to the Arab nations. Paper making was also brought to India by Chinese monks who traveled there searching for the Buddhist sutras. And through the Arabs, paper making skills were learned and mastered by the Africans and the Europeans. Paper making skills and paper then became widespread all across the globe.

With the invention of a cheap and easy writing surface, it meant that ideas, teachings and philosophies can now be easily passed on to other people. Education became a much easier task and communication with people from a distance is now simpler. The use of paper changed the way people taught and learned. It also promoted and hastened the progress of civilization and culture, and literature.

The Invention of Paper

The technique of paper making had gone through different processes being refined and perfected, from its humble beginnings in the Eastern Han Dynasty to the factory made and mass produced product it is today. It truly is one of the greatest inventions made by man.

33

THE INVENTION OF
THE INTERNET

Unlike technologies such as the light bulb or the telephone, the Internet has no single "inventor." Instead, it has evolved over time. The Internet got its start in the United States more than 50 years ago as a government weapon in the Cold War. For years, scientists and researchers used it to communicate and share data with one another. Today, we use the Internet for almost everything, and for many people it would be impossible to imagine life without it.

THE SPUTNIK SCARE

On October 4, 1957, the Soviet Union launched the world's first manmade satellite into orbit. The satellite, known as

The Invention of The Internet

Sputnik, did not do much: It tumbled aimlessly around in outer space, sending blips and bleeps from its radio transmitters as it circled the Earth. Still, to many Americans, the beach-ball-sized Sputnik was proof of something alarming: While the brightest scientists and engineers in the United States had been designing bigger cars and better television sets, it seemed, the Soviets had been focusing on less frivolous things—and they were going to win the Cold War because of it.

Did You Know?

Today, almost one-third of the world's 6.8 billion people use the Internet regularly.

After Sputnik's launch, many Americans began to think more seriously about science and technology. Schools added courses on subjects like chemistry, physics and calculus. Corporations took government grants and invested them in scientific research and development. And the federal government itself formed new agencies, such as the National Aeronautics and Space Administration (NASA) and the Department of Defense's Advanced Research Projects Agency (ARPA), to develop space-age technologies such as rockets, weapons and computers.

THE BIRTH OF THE ARPANET

Scientists and military experts were especially concerned about what might happen in the event of a Soviet attack on the nation's telephone system. Just one missile, they feared, could destroy the whole network of lines and wires that made efficient long-distance communication possible. In 1962, a scientist from M.I.T. and ARPA named J.C.R. Licklider proposed a solution to this problem: a "galactic network" of computers that could talk to one another. Such a network would enable government leaders to communicate even if the Soviets destroyed the telephone system.

In 1965, another M.I.T. scientist developed a way of sending information from one computer to another that he called "packet switching." Packet switching breaks data down into blocks, or packets, before sending it to its destination. That way, each packet can take its own route from place to place. Without packet switching, the government's computer network—now known as the ARPAnet—would have been just as vulnerable to enemy attacks as the phone system.

"LOGIN"

In 1969, ARPAnet delivered its first message: a "node-to-node" communication from one computer to another. (The first computer was located in a research lab at UCLA and the second was at Stanford; each one was the size of a small house.) The message—"LOGIN"—was short and simple, but it crashed the fledgling ARPA network anyway: The Stanford computer only received the note's first two letters.

THE NETWORK GROWS

By the end of 1969, just four computers were connected to the ARPAnet, but the network grew steadily during the 1970s. In 1971, it added the University of Hawaii's ALOHAnet, and two years later it added networks at London's University College and the Royal Radar Establishment in Norway. As packet-switched computer networks multiplied, however, it became more difficult for them to integrate into a single worldwide "Internet."

By the end of the 1970s, a computer scientist named Vinton Cerf had begun to solve this problem by developing a way for all of the computers on all of the world's mini-networks to communicate with one another. He called his invention "Transmission Control Protocol," or TCP. (Later, he added an additional protocol, known as "Internet Protocol." The acronym we use to refer to these today is TCP/IP.) One writer describes Cerf's protocol as "the 'handshake' that introduces distant and different computers to each other in a virtual space."

THE WORLD WIDE WEB

Cerf's protocol transformed the Internet into a worldwide network. Throughout the 1980s, researchers and scientists used it to send files and data from one computer to another. However, in 1991 the Internet changed again. That year, a computer programmer in Switzerland named Tim Berners-Lee introduced the World Wide Web: an Internet that was not simply a way to send files from one place to another but was itself a "web" of information that anyone on the Internet could retrieve. Berners-Lee created the Internet that we know today.

Since then, the Internet has changed in many ways. In 1992, a group of students and researchers at the University of Illinois developed a sophisticated browser that they called Mosaic. (It later became Netscape.) Mosaic offered a user-friendly way to search the Web: It allowed users to see words and pictures on the same page for the first time and to navigate using scrollbars and clickable links. That same year, Congress decided that the Web could be used for commercial purposes. As a result, companies of all kinds hurried to set up websites of their own, and e-commerce entrepreneurs began to use the Internet to sell goods directly to customers. More recently, social networking sites like Facebook have become a popular way for people of all ages to stay connected.

http://www.history.com/topics/inventions/invention-of-the-internet

34

THE INVENTION OF THE COMPUTER

Alan Turing and others

What is a Computer?

The machine which is sitting in front of you.

The machine which can draw graphics, set up your modem, decipher your PGP, do typography, refresh your screen, monitor your keyboard, manage the performance of all these in synchrony... and do all of these through a single principle: reading programs placed in its storage.

But the meaning of the word has changed in time. In the 1930s and 1940s "a computer" still meant aperson doing calculations. There is a nice historical example of this usage here. So to indicate a machinedoing calculations you would

The Invention of The Computer

say "automatic computer". In the 1960s people still talked about the digital computer as opposed to the analog computer.

But nowadays, I think it is better to reserve the word "computer " for the type of machine which has swept everything else away in its path: the computer on which you are reading this page, the digital computer with "internally stored modifiable program. "

The world's computer industries now make billions out of manufacturing better and better versions of Turing's universal machine. But Alan Turing himself never made a point of saying he was first with the idea. And his earnings were always modest. Picture from a Japanese comic book of Turing's story.

So I wouldn't call Charles Babbage's 1840s Analytical Engine the design for a computer. It didn't incorporate the vital idea which is now exploited by the computer in the modern sense, the idea of storing programs in the same form as data and intermediate working. His machine was designed to store programs on cards, while the working was to be done by mechanical cogs and wheels.

There were other differences — he did not have electronics or even electricity, and he still thought in base-10 arithmetic. But more fundamental is the rigid separation of instructions and data in Babbage's thought.

Charles Babbage, 1791-1871

- Charles Babbage (Wikipedia)

- Virtual Museum of Computing

- The completion of the Difference Engine in the Science Museum, London

- Home page maintained by his biographer A. Hyman.

A hundred years later, in the early 1940s, electromagnetic relays could be used instead of gearwheels. But no-one had advanced on Babbage's principle. Builders of large calculators might put the program on a roll of punched paper rather than cards, but the idea was the same: you built machinery to do arithmetic, and then you arranged for instructions coded in some other form, stored somewhere else, to make the machinery work.

To see how different this is from a computer, think of what happens when you want a new piece of software. You can download it from a remote source, and it is transmitted by the same means as email or any other form of data. You may apply an UnStuffIt or GZip program to it when it arrives, and this means operating on the program you have ordered. For filing, encoding, transmitting, copying, a program is no different from any other kind of data — it is just a sequence of electronic on-or-off states which lives on hard disk or RAM along with everything else.

The people who built big electromechanical calculators in the 1930s and 1940s didn't think of anything like this. I would call their machines near-computers, or pre-computers: they lacked the essential idea.

ENIAC

Even when they turned to electronics, builders of calculators still thought of programs as something quite different from numbers, and stored them in quite a different, inflexible, way. So the ENIAC, started in 1943, was a massive electronic calculating machine, but I would not call it a computer in the modern sense, though some people do. This page shows how it took a square root — incredibly inefficiently.

Colossus

The Colossus was also started in 1943 at Bletchley Park, heart of the British attack on German ciphers.

I wouldn't call it a computer either, though some people do: it was a machine specifically for breaking the "Fish" ciphers, although by 1945 the programming had become very sophisticated and flexible.

But the Colossus was crucial in showing Alan Turing the speed and reliability of electronics. It was also ahead of American technology, which only had the comparable ENIAC fully working in 1946, by which time its design was obsolete. (And the Colossus played a part in defeating Nazi Germany by reading Hitler's messages, whilst the ENIAC did nothing in the war effort.)

1996 saw the fiftieth anniversary of the ENIAC. The University of Pennsylvania and the Smithsonian made a great deal of it as the "birth of the Information Age". Vice-President Gore and other dignitaries were involved. Good for them. At Bletchley Park Museum, the Reconstruction of the Colossus had to come from the curator Tony Sale's individual efforts. Americans and Brits do things differently. Some things haven't changed in fifty years.

There is no doubt that Turing's war experience was what made it possible for him to turn his logical ideas into practical electronic machinery. This is a great irony of history which forms the central part of his story. He was the most civilian of people, an Anti-War protester of 1933.

He was very different in character from John von Neumann, who relished association with American military power. But von Neumann was on the winning side in the Second World War, whilst Turing was on the side that scraped through, proud but almost bankrupt.

The Internally Stored Modifiable Program

The breakthrough came through two sources in 1945:

- Alan Turing, on the basis of his own logical theory, and his knowledge of the Colossus.

- the EDVAC report, by John von Neumann, but gathering a great deal from ENIAC engineers Eckert and Mauchly. Download the EDVAC report in pdf form.

They both saw that the programs should be stored in just the same way as data. Simple, in retrospect, but not at all obvious at the time.

John von Neumann, 1903-1957

John von Neumann (originally Hungarian) was a major twentieth-century mathematician with work in many fields unrelated to computers.

- Wikipedia article

- MacTutor mathematical biography

- Tools for Thought, by Howard Rheingold

The EDVAC report became well known and well publicised, and is usually counted as the origin of the computer in the modern sense. It was dated 30 June 1945 — before Turing's report was written. It bore von Neumann's name alone, denying proper credit to Eckert and Mauchly who had already seen the feasibility of storing instructions internally in mercury delay lines.

Where did von Neumann get the idea?

He knew Turing at Princeton in 1937-38. By 1938 he certainly knew about Turing machines. See these source documents on what von Neumann knew of Turing, 1937-39. Many people have wondered how much this knowledge helped him to see how a general purpose computer should be designed.

The logician Martin Davis, who was involved in early computing himself, has written a book The Universal Computer, The Road from Leibniz to Turing. Martin Davis is clear that von Neumann gained a great deal from Turing's logical theory.

So who invented the computer?

There are many different views on which aspects of the modern computer are the most central or critical.

• Some people think that it's the idea of using electronics for calculating — in which case another American pioneer, Atanasoff, should be credited.

• Other people say it's getting a computer actually built and working. In that case it's either the tiny prototype at Manchester or the EDSAC at Cambridge, England (1949), that deserves greatest attention.

But I would say that in 1945 Alan Turing alone grasped everything that was to change computing completely after that date: above all he understood the universality inherent in the stored-program computer. He knew there could be just one machine for all tasks. He did not do so as an isolated dreamer, but as someone who knew about the practicability of large-scale electronics, with hands-on experience. From experience n codebreaking and mathematics he was also vividly aware of the scope of programs that could be run.

The idea of the universal machine was foreign to the world of 1945. Even ten years later, in 1956, the big chief of the electromagnetic relay calculator at Harvard, Howard Aiken, could write:

If it should turn out that the basic logics of a machine designed for the numerical solution of differential equations coincide with the logics of a machine intended to make bills for a department store, I would regard this as the most amazing coincidence that I have ever encountered.

But that is exactly how it has turned out. It is amazing, although we now have come to take it for granted. But it follows from the deep principle that Alan Turing saw in 1936: the Universal Turing Machine.

35

THE DEVELOPMENT OF THE MOTOR CAR

Who invented the world's very first car?

Who invented the first car? If we're talking about the first modern automobile, then it's Karl Benz in 1886. But long before him, there were strange forerunners to the today's cars,

The Development of The Motor Car

including toys for emperors, steam-powered artillery carriers, and clanking, creaking British buses.

Humans have possessed knowledge of the wheel for several thousand years, and we've been using animals as a source of transportation for nearly that long. So, in some sense, the earliest forerunners of the car date back to the earliest mists of our prehistory. But perhaps a more useful way of thinking of the car is anything that could reasonably be called an "automobile" - in other words, any vehicle capable of propelling itself. In that case, we're at most talking about 439 years of car history.

The Emperor's Toy

The very first car might well have been the invention of a Flemish missionary named Ferdinand Verbiest. Born in Flanders in 1623, Verbiest was an accomplished astronomer who left Europe for China in 1658. He helped to modernize the now outmoded Chinese astronomy using recent European innovations, and he was asked by the emperor to become the director of the newly refurbished Beijing Ancient Observatory. What's more, he spoke at least five languages fluently, wrote thirty books, was a skilled diplomat and mapmaker, and tutored the long-lived Kangxi Emperor in everything from mathematics to poetry. He was, even by the standards of the time, ridiculously accomplished.

But the reason why we're talking about Verbiest here is that he might - emphasis on might - have invented the world's first car. According to Verbiest's own text AstronomiaEuropea, he built a small, self-propelled vehicle. Steam technology was

The Development of The Motor Car

still in its infancy at the time, but Verbiest was able to build a rudimentary, ball-shaped boiler, which then forced steam towards a turbine that could turn the back wheels. Verbiest says the vehicle was meant to be a toy for the emperor.

Considering this is over 200 years before the construction of what's generally considered the first modern automobile, this is a remarkable achievement, but there are some pretty big caveats here. I said the car was small, and it was: about two feet long, far too tiny for any human to ride in it. It's also not at all clear whether the toy was ever built, or if it purely existed as a design in Verbiest's imagination.

We do know Verbiest's close relationship with the emperor gave him access to the finest metalworkers China had to offer, so it's not impossible that he built the toy. What we can say is this - Verbiest almost certainly designed what was effectively one of the earliest scale models of an automobile. (Although, if we're just talking about designs for cars, then Leonardo da Vinci has Verbiest beat by a good two hundred years. But Leonardo definitely didn't build his, so Verbiest has that on him.)

The First Engine

To some extent, 1672 might seem surprisingly recent for the first car ever. After all, we keep discovering far more ancient analogues for modern items, including everything from Babylonian museums to Roman fishtanks. So why haven't we discovered an ancient Egyptian car inside the pyramids, or even some medieval gadgetry that vaguely approximates an automobile?

Part of the reason why it took until 1672 for anyone to even build a toy version of a car was that there was just no need for them, and it wasn't really the sort of thing one could invent in one fell swoop. In World History of the Automobile, Erik Eckermann explains the basic problem:

The Development of The Motor Car

The wagon existed in its animal-drawn form for thousands of years before it was possible to make it self-propelled, literally "auto-mobile." In the process, motorized vehicles were far removed from the center of scientific and mechanical inquiry. From the end of the seventeenth century, existing vehicular technology was more than adequate to meet societal demands. In the age of absolute monarchs and mercantilism, it was more important to solve other engineering challenges that were difficult or impossible to achieve with conventional energy sources such as muscle, wind, or water power.

And what were these more important engineering challenges? As Eckermann explains, "the fountains and water displays of baroque gardens" were a higher priority for inventors and scholars than was the creation of a self-propelling vehicle.

While no one was really tackling this subject directly, the legendary Dutch scientist Christian Huygens did take a crucial step towards the car in 1673, one year after Verbiest reputedly began work on his toy for the emperor of China.

Huygens built upon previous experiments by other scientists to create a simple engine powered by, awesomely enough, gunpowder. By exploding the material inside a cylinder, Huygens was able to create a vacuum, which in turn forced a piston to move down the cylinder. This created work, making it effectively the earliest recognizable forerunner of the internal combustion engine. And, for his part, Huygens immediately recognized the engine's potential as a power source for land and water vehicles alike, but his engine was far too primitive to be of much use in that direction.

Cugnot's Car

The 1700s were dominated by various inventors working to perfect the steam engine - Thomas Newcomen and James Watt are probably the most famous of these, but there were many more. But the first person to take a steam engine and place it on a full-sized vehicle was probably a Frenchman named Nicolas-Joseph Cugnot, who between 1769 and 1771 built a steam-powered automobile more than thirty years before the railway's first steam locomotive.

Cugnot's design was, to put it mildly, unique. The contraption weighed about 2.5 tons, had two big wheels in the back and a single thick central wheel at the front, and could seat four people. The boiler was placed well out in the front, which made the vehicle even more fiendishly difficult to control. While its top speed was meant to be about five miles per hour, it never even got close to that fast in practice.

Opinion is somewhat divided over how well the thing actually worked - in fairness, various government ministers were supposed to be impressed with the initial trials - although most agree that it had poor weight distribution and so was unable to handle even moderately rough terrain. Since its intended purpose was as a transport for heavy artillery on the battlefield, that has to be considered a drawback.

One story says that the second of Cugnot's two vehicles crashed into a wall in 1771, which might make it the first ever automobile accident. It's a good story, but unfortunately no one wrote about it until 1801, some thirty years later, which makes it rather more likely that this was just a bit of folklore. Either way, here's a rather awesome reconstruction of the crash, completely with ludicrously over-the-top reaction shots.

The Steam Buses

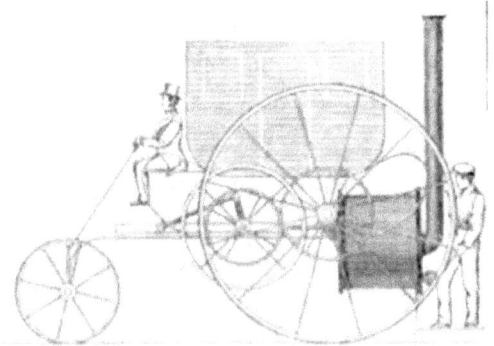

As France fell into the grips of revolution, Cugnot's work was largely forgotten, and the next big innovations in automobile technology came in Britain. Over the next several decades, various inventors worked on steam carriages, which resembled a cross between buses and rail locomotives. William Murdoch created a working model of one of these in 1784, but it wouldn't be until the beginning of the 19th century that Richard Trevithick was able to get a full-sized vehicle on the road.

Steam-powered mass transit had some limited success in the opening years of the 1800s, but it wasn't until the 1820s and 1830s that steam buses began gaining some measure of popularity with the British public. Further technological innovations in this early form of road-based mass transit including better brakes, a more advanced transmission, and improved steering.

But, as Erik Eckermann explains, the drawbacks still far outweighed the advantages of this new technology:

The Development of The Motor Car

It was apparent that the technology was not yet fully developed, and this new means of transportation did not yet enjoy favorable public opinion. Crankshafts snapped, lines leaked, chains broke, and boilers exploded. Engine vibrations (which, unlike stationary installations, could not be overcome by mounting on a solid foundation, the pungent odor of burnt oil, and flying soot and coal dust soon drove the traveling public back to the old standby, the horse-drawn stage, or another new invention, the railway and its rapidly growing network of track.

The steam buses proved to be something of a dead end, and engineers turned their attention to traction engines, which were slower, more stable machines that were basically just steam locomotives adapted for use on land. This was a move away from the line of innovation that would eventually lead to the car, but even these proved too raucous for the public at large. The Locomotive Act of 1865 said no land vehicle could travel faster than 4 miles per hour, and that all such vehicles had to be preceded by a man waving a red flag and blowing a horn. This was not, as you might imagine, the automotive industry's finest hour.

Other Curiosities

The Development of The Motor Car

There were several other attempts to build self-propelled vehicles, but none of them ever quite made that big leap to become the first practical automobile. An American inventor named Oliver Evans built the "OruktorAmphibolis", a steam-powered dredging device that became more powerful and elaborate with each subsequent retelling, in part because Evans felt he never got proper credit for his engineering prowess. At this point, it's difficult to say with certainty exactly what the OruktorAmphibolis was actually capable of.

Russian inventor Ivan Kulibin came other with a steam-powered vehicle in the 1780s, and it featured plenty of modern automotive hallmarks, including brakes, gearbox, flywheel, and bearing. The problem is that, though it did have a steam engine component, it still required human peddling to operate, so it can't really be considered an automobile.

While steam remained the main focus of inventors in search of a practical automobile, the results remained difficult to control and incapable of reaching speeds much over about five miles per hour. (In fairness, subsequent innovations in the late 1800s and early 1900s did result in actually practical steam cars.) The internal combustion engine provided the pathway to the first modern automobiles, with Karl Benz generally getting the credit for the first successful invention in 1886.

But now we're starting to cross over into the modern history of automobiles, so this is where I will stop. Here's to all the crazy forerunners of our modern marvel, be they Flemish polymath toymakers, Frenchmen crashing into walls, Dutchmen building engines out of gunpowder, or Brits crowding themselves onto noisy, supremely dangerous steam buses. All these innovators offer a very clear lesson: if you're going to fail to invent the automobile, at least fail with style.

36

The Invention of Radio

Guglielmo Marconi

Radio is indebted to two other discoveries for its birth-telephone and telegraph. These three technologies are very closely related. Radio started as wireless telegraphy. And it all began with invention of radio waves, which have capacity to send out speech, music, picture and all other data through air. A range of devices like radio, cordless phones, microwave TV broadcasts work with the help of electromagnetic waves.

During 1860s, James Maxwell, a Scottish physicist forecasted presence of radio waves. And in year 1886, Heinrich Hertz showcased projection of swift variation of the electric current into space in form of radio waves.

The Invention of Radio

Twenty years after invention of telephone, music was set down on telephone line and Guglielmo Marconi was responsible for the radio signals. This Italian discoverer demonstrated radio communication's feasibility. Fascinated by Hertz's discovery of the radio waves, he realized that it could be used for receiving and sending the telegraph messages, referring to it as wireless telegraphs.

His earliest radio transmissions, the coded signals transmitted to only a mile far in 1896. Marconi then recognized its high potential and offered the discovery to Italian Government that had turned it down. He then realized a patent and experimented further after moving to England. In 1898, he flashed the results of Kingstown Regatta to Dublin Newspaper's office, making first ever public broadcast of sports event. The following year, he opened his radio factory in Essex, thus establishing link between France and Britain. He then established link with USA in year 1901. And Marconi shared Nobel Prize in Physics for wireless telegraph in year 1909.

But his wireless telegraph only transmitted signals. Voice in radio came in the 1921. Soon after, in 1922, he introduced short wave transmissions. Marconi however was not the first one to invent radio. Nikola Tesla who moved to US in 1884, launched radio's theoretical model prior to Marconi. In 1915, Tesla tried to acquire court's injunction against Marconi. And in year 1943, Supreme Court US reviewed decision. And due to this Tesla was acknowledged as inventor of radio even when he did ever build working radio.

J.C. Bose was another claimant to throne of radio inventors. He showcased radio transmission to British Governor General in 1896 at Calcutta. The transmission was for a distance of around 3 miles. His instruments, Mercury Coherer attached to telephone detector, are still showcased Calcutta University's Science College.

Bose had taken care of Hertz's problem of not being able to penetrate through water, mountains or walls. Marconi's Coherer is known to be exact copy of Bose's Coherer. Initially, Bose was reluctant in applying for patent because he believed in free flow of inventions in the field of science. But eventually due to persuasion by his American friends he had applied for patent in 1901. US patent was awarded to him in year 1904.

There has been tremendous growth of the radio over the years. Transmitters earlier were known as spark gap machines. It was established for ship-to-ship and ship-to-shore communication. The communication was just confined to two points then and was not public broadcasting as it is today. Wireless signals demonstrated effective communication for the purpose of rescue in case of sea disasters. Range of ocean liners installed the wireless equipments and in 1899, US Army established the wireless communication. Just after two years, Navy adopted wireless system and it was relief as Navy had been using homing pigeons and visual signaling for communication.

Radiotelegraph services were instituted in Hawaiian Islands in 1901. Marconi station situated in Massachusetts carried greetings between King Edward VII and Theodore Roosevelt. In year 1905, Port Arthur's naval battle was also reported over wireless and US weather department tried radiotelegraphy for speeding notice weather condition.

Eventually radio transmitters were improved. Overseas radiotelegraph services were slowly developed, basically because early transmitter discharged the electricity between electrodes and within circuit causing high interference. DeForest and Alexanderson alternator took care of many such technical issues.

Lee Deforest was inventor of space telegraphy, Audion and triode amplifier. In early 1990s, delicate and effective detector of the electromagnetic radiation was needed for developing the radio further. And Lee Deforest discovered the detector. He was the first person to use term 'radio'. His work resulted in discovery of AM radio that capably broadcasted various radio stations which early gap transmitters did not allow.

Since that time, there has been no looking back. Radio has now become a popular medium of portable entertainment. In 21stcentury, technological advancements have given birth to internet radio. Satellite radio is also recent development in the field. One can listen to various international radio stations without any hassles. Besides all these latest editions, Ham radio would be next big thing. This technology is gearing up to hit the market soon. Radio lovers have a reason to rejoice as there is a lot in store for them.

http://www.engineersgarage.com/invention-stories/radio-history

37

INVENTION OF THE TELEPHONE

Alexander Graham Bell

When the word "inventor" is mentioned, Alexander Graham Bell, creator of the telephone, is undoubtedly one of the first names that springs to mind.

Bell was born on March 3, 1847, in Edinburgh, Scotland, and educated at the universities of Edinburgh and London. He immigrated to Canada in 1870 and to the United States in 1871. He was an early student of sound and speech, inspired, perhaps, by the fact that his mother, Eliza, was almost totally deaf and his father, Melville, developed the first international phonetic alphabet. In his early 20s Bell himself taught deaf children to speak and gave speech lessons at schools in his community.

As a boy, Bell built a speaking robot, and found that he could touch his dog's throat in ways that seemed to form his barks and growls into words. Once, he successfully obtained a human ear from a medical school, which he used to conduct experiments tracing sound patterns. Bell was also a gifted pianist, who learned to discriminate pitch very well. As a teenager, he noticed that a chord struck on a piano in one room would be echoed by a piano in another room. He realized that chords could be transmitted through the air, vibrating at the other end at exactly the same pitch.

With this discovery, Bell set out to develop a multiple telegraph, using Morse code to convey several messages simultaneously, each at a different pitch. He knew his greatest challenge would be finding a way to convey pitch across a wire. He ascertained, eventually, that this could be accomplished by reproducing sound waves in a continuous, undulating current. That's when he realized that this could also apply to human speech, which is composed of many complex sound vibrations.

In 1875, Bell developed his first version of what came to be known as the telephone. He received a patent for it on March 7, 1876, just after his 29th birthday. Five days later, on March 12, he tested his device, speaking into the phone to his associate, Thomas Watson, when he said, "Mr. Watson, come here. I want to see you."

Bell first demonstrated his most famous invention on June 25, 1876 at the Centennial Exhibition in Philadelphia. There, he showed that the sound of the human voice could be reproduced, which confirmed his theory that speech patterns can be made to change the intensity of an electrical current.

A year after Bell's initial public demonstration, he placed the world's first phone call over telegraph wires between two towns in Ontario, Canada -- a span of eight miles. Just two months later, the long-distance reach of telephone technology was expanded to 143 miles. Today, of course, telephone calls may be placed to virtually any location around the globe. The Bell Telephone Company was established in 1877 to bring telephones to the masses. The company provided the foundation for today's telecommunications industry.

While Bell is best known for his telephone invention, he worked on hundreds of projects throughout his life and received a number of patents in various fields.

In 1880, Bell, patented the photophone, in which his telephone principle was applied to transmit words on a beam of light. This has been recognized as the first wireless transmission of speech. Not until more than a century later would this idea have any widespread use: the principles behind the process enabled the development of what we know today as the cellular phone.

Invention of the Telephone

Bell was also an aviation enthusiast. He worked on designs for airplanes, kites and helicopters with members of the Aerial Experiment Association. In 1909, Bell's Silver Dart airplane few for a half mile in Baddeck, Nova Scotia, six years after the Wright Brothers took their first flight in North Carolina. Later Bell developed the tetrahedron while he worked on the design for a kite that could carry a man. The figure, made up of four equilateral triangles, is one of nature's most stable structures and forms the basis for many modern bridges and towers. At the age of 75, Bell received a patent on one of the fastest watercraft in the world, the HD-4.

To sum up his approach to invention, Bell once said, "Leave the beaten track behind occasionally and dive into the woods. Every time you do you will be certain to find something that you have never seen before. Follow it up, explore all around it, and before you know it, you will have something worth thinking about to occupy your mind."

Bell's notebooks are still available for public consultation. Researchers believe his early ideas may still hold clues that can help provide the solutions for modern technological problems.

http://web.mit.edu/invent/iow/graham_bell.html

38

THE INVENTION OF THE TELEVISION

John Logie Baird

"No great man lives in vain. The history of the world is but the biography of great men." -Thomas Carlyle 1795-1881

Television, a telecommunication medium for transmitting and receiving moving images was first demonstrated publicly by Scottish inventor John Logie Baird in 1925. Baird's scanning disk produced an image of 30 lines resolution, just enough to discern a human face, from a double spiral of lenses. Baird invented television. It wasn't the television that was taken up universally as the standard format for TV broadcasts throughout the 1940s and beyond. And that is why, even before television development of the 40s was little more than a fading memory, it became popular to dismiss Baird's contribution as "...nothing more than that of a romantic who was destined to nothing more than failure due to the fact that his mechanical system was replaced by electronic television." (from a 1952 biography of Baird subtitled "The Romance and Tragedy of the Pioneer of Television" by Sydney Moseley). But Baird did something that nobody had managed before. He transmitted a live moving picture from one location to another. That was what he invented and that was television.

In 1957 an attempt to convert the Baird family home in Scotland to a public museum of television was thwarted by some "experts" who said bluntly, "Baird did not invent television." In a book on television by Francis Wheen, the author seemingly went out of his way to discredit the pioneering work of Baird by quoting negative remarks and dismissing most of his groundbreaking work as feats of "one-upmanship", whilst disregarding any of Baird's recognised achievements. And even in 1991 in his book "Setmakers -A History of the Radio and Television Industry", author Keith Geddes stated, ..."apart from stimulating public interest in television it (Baird's low-definition television) contributed nothing to the high definition systems that succeeded it."

To dismiss Baird's contributions to the development of modern day television does a great disservice to one of the most brilliantly gifted inventors of our time. As you will read for yourself, Baird at times overcame insurmountable odds, ill health and poverty to achieve what many of his contemporaries regarded as the impossible. However, there was also another side to Baird that contradicts many of the so-called "established facts" about the man and his work. Even if only a small part of this other side of Baird's life story is to be believed, then the reader, by the end of this biography, should be in no doubt whatsoever that John Logie Baird's contribution to television, as well as modern day broadcasting technology has, for far too long, been seriously underrated.

Research for this biography includes excerpts from the first full biography of Baird, "Baird of Television" written by Ronald F. Tiltman, which was written with the cooperation of John Logie Baird himself (as well as some Baird family members) and featured contributions from many of the people who witnessed his experiments less than a decade before the book's publication in 1933. This book is no doubt the basis for many reference works on both the life of Baird and his work in the development of television. Another interesting source is a 1986 publication entitled "The Secret Life of John Logie Baird" by Tom McArthur and Peter Waddell. A detailed account of Baird's experiments into television and beyond. The book dispels many of the previously held myths surrounding Baird's work as well as disputing previously established "facts" about when, where and how Baird first realised his dream of creating "true" television. And by unearthing previously unknown details about the inventor and his secret work beyond the development of television, offers convincing evidence to support its claims.

It will also be seen that Baird didn't do himself any favours when it came to cementing his reputation and evidencing his own contributions. Much of his work was kept secret and dates were changed in order to hide the projects he was actually working on which has left inconsistencies in important dates and confusion with regards to certain "discoveries."

In the preface to his 1933 publication, Mr Tiltman wrote, "Baird was the first man in the world to achieve television, the first man to commercialise television. He placed British television in the van of world progress and, in my opinion, has maintained its pre-eminent position up till now. Who will deny this pioneer the status which is his by right of his accomplishments?"

http://www.teletronic.co.uk/john_logie_baird.htm

39

INVENTION OF PAPER MONEY

The Chinese

The history of paper money is interesting not only from the idea and technology of printing, but also from the perspective of trading with a commodity that in itself has no intrinsic value. Clearly the issues of paper currency must inspire confidence for trading something of worth for items of no specific worth, and with the potential to be abused by the issuer as a way to increase the supply and control of items of value, thus creating inflation.

For much of its history, China used gold, silver and silk for large sums, and bronze for everyday transactions. The notion of using paper as money is almost as old as paper itself. The

The Invention of Paper Money

first paper banknotes appeared in China about 806 AD. An early use of paper was for letters of credit transferred over large distances, a practice which the government quickly took over from private concerns. The Chinese, with their great gift for pragmatism, labelled this practice "flying money". The printed notes were normally military scrip or other emergency measures issued in dire circumstances, but for the most part these notes disappeared quickly. The first real use of a paper money system was in Szechwan province, an isolated area subject to frequent copper shortages (which is a component of bronze). It had reverted to an iron currency of coins, and paper was a welcome option. Iron banks sprang up to facilitate the trade, and the government was quick to take over the profitable enterprise. Amazingly, the Chinese only used paper money on any meaningful scale for about 300 years of a 400 year period between 1050 and 1450, overlapping the Song, Yuan (Mongol), and Ming dynasties.

The Song dynasty was the first to issue true paper money in 1023, and it did so at first cautiously, issuing small amounts, used in a limited area, and good for a specific time period. The notes would be redeemed after three year's service, to be replaced by new notes for a 3% service charge, an efficient way for the government to make money.

The most famous Chinese issuer of paper money was Kublai Khan, the Mongol who ruled the Chinese empire in the 13th century. Kublai Khan established currency credibility by decreeing that his paper money must be accepted by traders on pain of death. As further enforcment of his mandate, he confiscated all gold and silver, even if it was brought in by foreign traders. Marco Polo was impressed by the efficiency of the Chinese system, as he chronicles in his The Travels of Marco Polo (Il Milione).

"All these pieces of paper are issued with as much solemnity and authority as if they were of pure gold or silver; and on every piece a variety of officials, whose duty it is, have to write their names, and to put their seals. And when all is prepared duly, the chief officer deputed by the Khan smears the seal entrusted to him with vermilion, and impresses it on the paper, so that the form of the seal remains imprinted upon it in red; the money is then authentic. Anyone forging it would be punished with death. And the Khan causes every year to be made such a vast quantity of this money, which costs him nothing, that it must equal in amount all the treasure of the world."

As is to be expected, paper money did not succeed everywhere. In Persia, its forcible introduction in 1294 led to a total collapse of trade. By the 15th century even China had more or less given up paper money. Over this period, paper notes were issued irresponsibly, to the point that their value rapidly depreciated and inflation soared. Then beginning in 1455, the use of paper money in China disappeared for several hundred years. This was still many years before paper currency would reappear in Europe, and three centuries before it was considered common.

Western civilization had minted precious metal objects and coins for trade since about 500 BC. Devaluation and inflation often destroyed a monetary system. Banking systems were cyclic with nations and rulers, and the need to transfer large sums of money to finance the Crusades provided a stimulus to the re-emergence of banking in western Europe. In Europe, the first issuer of paper money was Sweden, where in 1661 Johan Palmstruch's Stockholm Banco introduced the first banknotes. Unfortunately, the bank subsequently overextended itself and had to call in government aid. Despite this example, other European countries soon followed the Swedish lead. In 1694 the Bank of England was established and was soon printing "running cash notes".

http://www.computersmiths.com/chineseinvention/papermoney.htm

40

The Invention of the Aeroplane

Early Efforts of Flight

Around 400 BC - China

The discovery of the kite that could fly in the air by the Chinese started humans thinking about flying. Kites were used by the Chinese in religious ceremonies. They built many colorful kites for fun, also. More sophisticated kites were used to test weather conditions. Kites have been important to the invention of flight as they were the forerunner to balloons and gliders.

Humans try to fly like birds

For many centuries, humans have tried to fly just like the birds. Wings made of feathers or light weight wood have been attached to arms to test their ability to fly. The results were often disastrous as the muscles of the human arms are not like a birds and cannot move with the strength of a bird.

Hero and the Aeolipile

Hero of Alexandria's Aeolipile

Aeolipile The ancient Greek engineer, Hero of Alexandria, worked with air pressure and steam to create sources of power. One experiment that he developed was the aeolipile which used jets of steam to create rotary motion.

Hero mounted a sphere on top of a water kettle. A fire below the kettle turned the water into steam, and the gas traveled through pipes to the sphere. Two L-shaped tubes on opposite sides of the sphere allowed the gas to escape, which gave a thrust to the sphere that caused it to rotate.

THE INVENTION ON THE AERIOPLANE

1485 Leonardo da Vinci - The Ornithopter

Leonardo da Vinci's Ornithopter Leonardo da Vinci made the first real studies of flight in the 1480's. He had over 100 drawings that illustrated his theories on flight.

The Ornithopter flying machine was never actually created. It was a design that Leonardo da Vinci created to show how man could fly. The modern day helicopter is based on this concept.

1783 - Joseph and Jacques Montgolfier- the First Hot Air Balloon

The Invention on The Aerioplane

One of The Montgolfier's Balloons   The brothers, Joseph Michel and Jacques Etienne Montgolfier, were inventors of the first hot air balloon. They used the smoke from a fire to blow hot air into a silk bag. The silk bag was attached to a basket. The hot air then rose and allowed the balloon to be lighter-than-air.

In 1783, the first passengers in the colorful balloon were a sheep, rooster and duck. It climbed to a height of about 6,000 feet and traveled more than 1 mile.

After this first success, the brothers began to send men up in balloons. The first manned flight was on November 21, 1783, the passengers were Jean-Francois Pilatre de Rozier and Francois Laurent.

1799 - 1850's - George Cayley

One Version of a Glider   George Cayley worked to discover a way that man could fly. He designed many different versions of gliders that used the movements of the body to control. A young boy, whose name is not known, was the first to fly one of his gliders.

The Invention on The Aerioplane

Over 50 years he made improvements to the gliders. He changed the shape of the wings so that the air would flow over the wings correctly. He designed a tail for the gliders to help with the stability. He tried a biplane design to add strength to the glider. He also recognized that there would be a need for power if the flight was to be in the air for a long time.

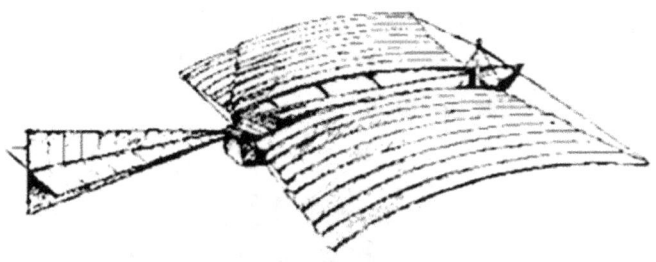

One of the many drawings of gliders

Cayley wrote On Ariel Navigation which shows that a fixed-wing aircraft with a power system for propulsion and a tail to assist in the control of the airplane would be the best way to allow man to fly.

19th And 20th Century Efforts

1891 Otto Lilienthal

One of Lilienthal's Gliders German engineer, Otto Lilienthal, studied aerodynamics and worked to design a glider that would fly. He was the first person to design a glider that could fly a person and was able to fly long distances.

The Invention on The Aerioplane

He was fascinated by the idea of flight. Based on his studies of birds and how they fly, he wrote a book on aerodynamics that was published in 1889 and this text was used by the Wright Brothers as the basis for their designs.

After more than 2500 flights, he was killed when he lost control because of a sudden strong wind and crashed into the ground.

Lilienthal's Glider in Flight

1891 Samuel P. Langley

Langley's Aerodrome

Samuel Langley was an astronomer, who realized that power was needed to help man fly. He built a model of a plane, which he called an aerodrome, that included a steam-powered engine.

The Invention on The Aerioplane

In 1891, his model flew for 3/4s of a mile before running out of fuel.

Langley received a $50,000 grant to build a full sized aerodrome. It was too heavy to fly and it crashed. He was very disappointed. He gave up trying to fly. His major contributions to flight involved attempts at adding a power plant to a glider. He was also well known as the director of the Smithsonian Institute in Washington, DC

Model of Langley Aerodrome

1894 Octave Chanute

Octave Chanute published Progress in Flying Machines in 1894. It gathered and analyzed all the technical knowledge that he could find about aviation accomplishments. It included all of the world's aviation pioneers. The Wright Brothers used this book as a basis for much of their experiments. Chanute was also in contact with the Wright Brothers and often commented on their technical progress.

The invention on The Aerioplane

Orville and Wilbur Wright and the First Airplane

Orville and Wilbur Wright were very deliberate in their quest for flight. First, they read about all the early developments of flight. They decided to make "a small contribution" to the study of flight control by twisting their wings in flight. Then they began to test their ideas with a kite. They learned about how the wind would help with the flight and how it could affect the surfaces once up in the air.

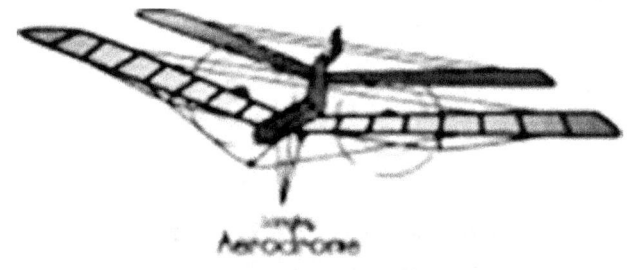

A Drawing of a Wright Brothers Glider (1900) The next step was to test the shapes of gliders much like George Cayley did when he was testing the many different shapes that would fly. They spent three years testing and learning about how gliders could be controlled at Kitty Hawk, North Carolina.

Picture of the actual 12 horsepower engine used in flight
They designed and used a wind tunnel to test the shapes of the wings and the tails of the gliders. In 1902, with a perfected glider shape, they turned their attention to how to create a propulsion system that would create the thrust needed to fly.

The Invention on The Aerioplane

The early engine that they designed generated almost 12 horsepower. That's the same power as two hand-propelled lawn mower engines!

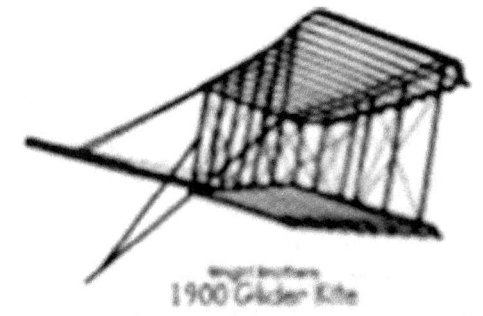

1900 Glider Kite

The Wright Brother's Flyer

The "Flyer" lifted from level ground to the north of Big Kill Devil Hill, North Carolina, at 10:35 a.m., on December 17, 1903. Orville piloted the plane which weighed about six hundred pounds.

The Invention on The Aerioplane

Actual Flight of The Flyer at Kitty Hawk

The first heavier-than-air flight traveled one hundred twenty feet in twelve seconds. The two brothers took turns flying that day with the fourth and last flight covering 850 feet in 59 seconds. But the Flyer was unstable and very hard to control.

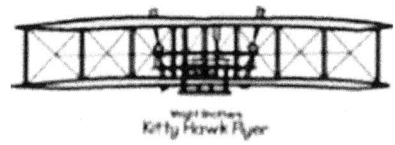

The brothers returned to Dayton, Ohio, where they worked for two more years perfecting their design. Finally, on October 5, 1905, Wilbur piloted the Flyer III for 39 minutes and about 24 miles of circles around Huffman Prairie. He flew the first practical airplane until it ran out of gas.

Humankind was now able to fly! During the next century, many new airplanes and engines were developed to help transport

THE INVENTION ON THE AERIOPLANE

people, luggage, cargo, military personnel and weapons. The 20th century's advances were all based on these first flights by the American Brothers from Ohio

41

THE DEVELOPMENT OF CEMENT

Many millennia ago, man was creating similar products to concrete and cement by pulverizing limestone or gypsum and then burning it. When you look at many of the newer mixes that use lime, it often refers to the same base material. This powder was then blended with water and sand to create a form of mortar that could be used to build with.

From Ancient Mortar to Roman Roads

Throughout the course of time, differing forms of concrete and cement have been used. For example, the Nabataeans (early Bedouins) were creating a form of this material for flooring to build their homes and cisterns from. The Chinese used sticky rice to build the Great Wall with its wide top used

THE DEVELOPMENT OF CEMENT

for transportation and defense. Even the Egyptians had their own form of mortar used to build the pyramids. However, the Romans truly made use of this invention to aid in building roads and bridges aimed at improving travel.

The Advent of Portland Cement

The invention of what would come to be known as the Portland blend occurred in 1824. This new blend was issued a patent for its unique mixture. Master mason and bricklayer Joseph Aspdin of Leeds in England developed it by measuring out the proportions of limestone and clay carefully before burning them to create the clinker. This clinker would then be ground in a fine powder used to make the final product. Aspdin named his new product Portland because the color of the final product looked similar to that of the stone which was being quarried and brought in from the Isle of Portland just of the coast.

DID YOU KNOW?

When the first highways were poured using cement concrete, the joints were formed during pouring. This created rough roads that could be hazardous to most forms of transportation. The invention of the concrete saw changed this because the joints could be cut after the mixture had begun to harden. This made them much smaller and smoother.

What Does Modern Portland Cement Concrete Contain?

For centuries, the burned lime form of concrete cement was used throughout the world. It was effective, strong, and easily produced. In fact, many buildings, such as the Coliseum and the Pantheon, were built using this mix. After many centuries, the fact they are still standing is a testament to the strength of the limestone blend. In many cases, the rock erodes instead of the mortar. Modern Portland concrete cement still contains limestone but, instead of being burned, it is superheated in a kiln along with clay. This forms a type of clinker that is then ground into powder to which gypsum is added.

The First All Concrete Cement Road in the US

While concrete and cement had a very long history of being used in the construction of buildings, it was not until 1891 that it was used to build a road. George Bartholomew poured this stretch of road, Court Street in Bellefontaine, Ohio. The quality of the material used to pour this road tested at over 8,000 pounds per square inch. This is approximately twice the strength of today's mixes. It may also be why the road is still in place and being used today without having undergone any major repairs or improvements.

Concrete Cement and the Highways

In 1913, the first section of highway in the United States was poured using concrete cement to make the pavement. It covered 24 miles and was 5 inches thick, spanning a width of 9 feet. The highway was just outside of Pine Bluff in Arkansas. One year later, there were over 2,300 miles of highway made from this material. By 1919, Oregon had become the first state to charge a tax on fuels to help fund the cost of installing new highways. In 1930, Pennsylvania began construction of the

THE DEVELOPMENT OF CEMENT

Pennsylvania Turnpike. This would be the first intercity toll road in the country and was made entirely of this material.

The history of concrete and cement goes back many centuries to ancient Egypt and China. In ancient Rome, builders first began to construct roads and bridges using this incredibly strong material. The first road in the United States to be made from it was poured in 1891, and it still in use today. While the 1960s and 1970s are considered to be the peak years for the use of this material in road construction, it is still used in road construction around the world today.

http://invent.answers.com/transportation/the-history-behind-the-invention-of-concrete-and-cement

42

THE ABACUS

Why does the abacus exist?

It is difficult to imagine counting without numbers, but there was a time when written numbers did not exist. The earliest counting device was the human hand and its fingers, the feet and toes. Then, as even larger quantities (larger than ten human-fingers and toes could represent) were counted, various natural items like pebbles and twigs were used to help keep count.

Merchants who traded goods not only needed a way to count goods they bought and sold, but also to calculate the cost of those goods. Until numbers were invented, counting devices were used to make everyday calculations. The abacus is one of many counting devices invented to help count large numbers.

The difference between a counting board and an abacus

The Abacus

It is important to distinguish the early abacuses (or abaci) known as counting boards from the modern abaci. The counting board is a piece of wood, stone or metal with carved grooves or painted lines between which beads, pebbles or metal discs were moved. The abacus is a device, usually of wood (plastic, in recent times), having a frame that holds rods with freely-sliding beads mounted on them.

Both the abacus and the counting board are mechanical aids used for counting; they are not calculators in the sense we use the word today. The person operating the abacus performs calculations in their head and uses the abacus as a physical aid to keep track of the sums, the carrys, etc.

What did the first counting board look like?

The earliest counting boards are forever lost because of the perishable materials used in their construction. However, educated guesses can be made about their construction, based on early writings of Plutarch (a priest at the Oracle at Delphi) and others.

In outdoor markets of those times, the simplest counting board involved drawing lines in the sand with ones fingers or with a stylus, and placing pebbles between those lines as place-holders representing numbers (the spaces between 2 lines would represent the units 10s, 100s, etc.) Affluent citizens could afford small wooden tables having raised borders that were filled with sand (usually coloured blue or green). A benefit of these counting boards on tables, was that they could be moved without disturbing the calculation— the table could be picked up and carried indoors.

With the need for portable devices, wooden boards with grooves carved into the surface were then created and wooden markers (small discs) were used as place-holders. The wooden boards then gave way to even more more durable materials like marble and metal (bronze) used with stone or metal markers.

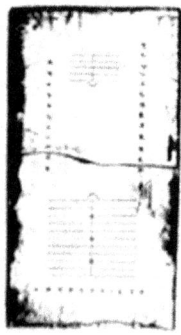

The Salamis Tablet: The oldest counting board, is made of marble. Photo from the National Museum of Epigraphy, Athens.

The Salamis Tablet

The oldest surviving counting board is the Salamis tablet (originally thought to be a gaming board), used by the Babylonians circa 300 B.C., discovered in 1846 on the island of Salamis.

It is a slab of white marble measuring 149cm in length, 75cm in width and 4.5cm thick, on which are 5 groups of markings.

The Abacus

In the center of the tablet are a set of 5 horizontal parallel lines divided equally by a perpendicular vertical line, capped with a semi-circle at the intersection of the bottom-most horizontal line and the vertical line.

Below these lines is a wide space with a horizontal crack dividing it. Below this crack is another group of eleven parallel lines, again divided into two sections by a line perpendicular to them but with the semi-circle at the top of the intersection; the third, sixth and ninth of these lines are marked with a cross where they intersect with the vertical line.

Three sets of Greek symbols (numbers from the acrophonic system) are arranged along the left, right and bottom edges of the tablet2.

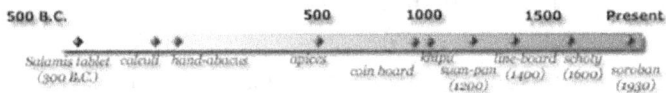

Evolution: The Abacus Through the Ages

The evolution of the abacus can be divided into three ages: Ancient Times, Middle Ages, and Modern Times. The time-line below traces the developing abacus from its beginnings circa 500 B.C., to the present.

Evolutionary Time-line: This time-line shows the evolution from the earliest counting board to the present day abacus. (Compared to the rate of progress in last one-thousand years, the progress during the first one-thousand years of civilization was rather slow).

Ancient Times

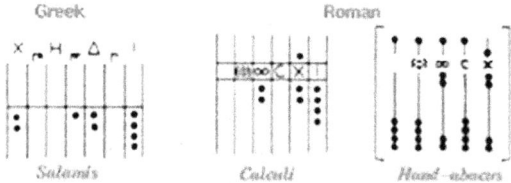

Ancient Times: The Salamis Tablet, the Roman Calculi and Hand-abacus are from the period c. 300 B.C to c. 500 A.D.

During Greek and Roman times, counting boards, like the Roman hand-abacus, that survive are constructed from stone and metal (as a point of reference, the Roman empire fell circa 500 A.D.).

The Middle Ages

Middle Ages: The Apices, the coin-board and the Line-board are from the period c. 5 A.D. to c. 1400 A.D.

Wood was the primary material from which counting boards were manufactured; the orientation of the beads switched from vertical to horizontal. As arithmetic (counting using written numbers) gained popularity in the latter part of the Middle Ages, the use of the abacus began to diminish in Europe.

Modern Times

Modern Times: The Suan-pan, the Soroban and the Schoty are from the period c. 1200 A.D to the present.

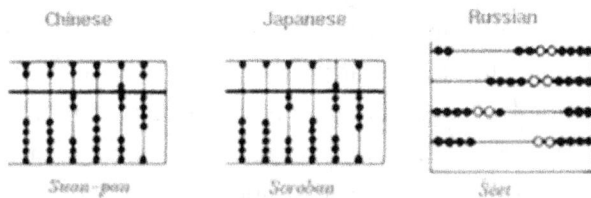

The abacus as we know it today, appeared (was chronicled) circa 1200 A.D. in China; in Chinese, it is called suan-pan. On each rod, this classic Chinese abacus has 2 beads on the upper deck and 5 on the lower deck; such an abacus is also referred to as a 2/5 abacus. The 2/5 style survived unchanged until about 1850 at which time the 1/5 (one bead on the top deck and five beads on the bottom deck) abacus appeared.

0, 1, 2, 3, 4, 5, 6, 7, 8, 9: The Hindu-Arabic Numbering System

In 1202, Leonardo of Pisa, also known as Fibonacci published Liber Abaci(Latin for The Book of Calculation) which illustrated, with examples that merchants could reference in their daily transactions, the superiority of calculations with

Arabic numbers over the Roman numbering system and counting boards. Fibonacci learned of the Arabic numbering system when he accompanied his father, a merchant, to various Arab ports in the Mediterranean Sea.

The comparative speed in which calculations with Arabic numbers were performed, was enough to obsolete counting boards in Western Europe.

Circa 1600 A.D., use and evolution of the Chinese 1/5 abacus was begun by the Japanese via Korea. In Japanese, the abacus is calledsoroban. The 1/4 abacus, a style preferred and still manufactured in Japan today, appeared circa 1930. The 1/5 models are rare today and 2/5 models are rare outside of China (excepting Chinese communities in North America and elsewhere).

It is thought that early Christians brought the abacus to the East (note that both the suan-pan and the Roman hand-abacus have a vertical orientation). Aspects of Roman culture could have been introduced to China as early as 166 A.D, during the Han Dynasty, as Roman emperor Antoninus Pius' embassies to China spread along the Silk Road.

There have been recent suggestions of a Mesoamerican (the Aztec civilization that existed in present day Mexico) abacus called the Nepohualtzitzin, circa 900-1000 A.D., where the counters were made from kernels of maize threaded through strings mounted on a wooden frame. There is also debate about the Incan Khipu— was it a three-dimensional binary calculator or a form of writing? (q.v.Talking Knots of the Incas).

The schoty, is a Russian abacus invented in the 17th century and still used today in some parts.

The Lee Kai-chen Abacus: Further refinement of the Chinese abacus c.

The Abacus Today

The image is a cover of a manual published in 1958 by Lee Kai-chen, the inventor of this "new" abacus designed with 4 decks (it combines two abaci; the top abacus is a small 1/4 sorobanand the bottom one is a 2/5 suan-pan).

The author claims that multiplication and division are easier using this modified abacus and includes instructions for determining square roots and cubic roots of numbers.

http://www.ee.ryerson.ca/~elf/abacus/history.html

43

The History of the Nail

Paul Fourshee

The lowly nail's history goes back several thousand years. While the nail has almost always been produced for fastening and joining, historically some other fairly imaginative applications have been made of this versatile product, such as mayhem and punishment.

Bronze nails, found in Egypt, have been dated 3400 BC. The Bible give us numerous references to nails, the most well known being the crucifixion of Christ. Of course we should not forget that model wife in Judges who in 1296 BC drove a nail into the temple of her husband while he was asleep, "so he died." (Thelma and Louise where is your imagination?)

Exactly what do we mean when we refer to nail sizes by "penny?" You're in good company if you have no idea.

With 2,200 varieties of nails being manufactured today and everyone using them from the hobbyist to the professional builder, one would think, if it is such a good idea, that somebody would know what the term "penny" means and who started it. At long last an answer to the question you never asked.

The term "penny", as it refers to nails, is thought to have originated in medieval England to describe the price of 100 nails. (e.g. 100 3-1/2" nails would cost 16 pence, while 100 2-1/2" nails could be bought for 6 pence.) This system of classifying nails by size according to price was in place by 1477 AD. The letter "d", which means penny, stands for the Latin name given to Roman Coins, Denarius.

The size of the nail is determined by measuring its length. Nails start at 2d, which is 1" in length, and range up to 60d which is 6" in length. From 2d to 16d the penny length increases by quarter inches. Above 16d, the size increases by half inches. Nails longer than 60d or shorter than 2d are described in inches or fractions thereof.

Just prior to the American Revolution, England was the largest manufacturer of nails in the world. Nails were virtually impossible to obtain in the American Colonies so it was quite common for families to have a small nail manufacturing setup in their homes by the fireplace. During bad weather and at night, entire families made nails not only for their own use but also for barter.

The History of the Nail

This was not a practice restricted to the lower classes, Thomas Jefferson was quite proud of his hand made nails. In a letter he wrote, "In our private pursuits it is a great advantage that every honest employment is deemed honorable. I am myself a nail maker." From the president to the pioneer, nail making was an important facet of life. Jefferson was among the first to purchase the newly invented nail-cutting machine in 1796 and produce nails for sale.

Such value was placed on nails that it was common practice, when moving, to burn one's home in order to retrieve them.

The invention of the nail cutting machine rapidly put the United States in front in the manufacturing of nails and has lead the world ever since.

In the 1850's several manufactures were established in New York which made wire nails. These machines were most likely imported from France. The earliest wire nails were not made for construction but for the manufacture of pocket book frames and cigar boxes. It was not until after the American War Between the States that wire nails began to gain acceptance in construction. Even through the 1890's many builders preferred using cut nails because of their holding power. It was well into the twentieth century before wire nails became the dominate type and only then because they were so much cheaper.

It is because of the tremendous holding power and hardness that cut nails are still used today for specific functions such as flooring nails, boat nails and masonry nails.

The Tremont Nail Company of Wareham, Massachusetts was

established in 1819 and has manufactured cut nails continuously under several owners and names ever since. This company, now owned by Maze Nails, still makes 20 different types of cut nails with 100 year old machines. Their nails are still packaged in 100 # wooden kegs.

Did you know that the holding power of common nails drops by half within two days after being driven? After about a month the holding power will increase slightly as the wood fibers straighten out and grip the nail.

Cement coated nails hold more securely than common nails but wet wood will loosen the cement coating in a matter of days. Threaded or ring shank nails loose their holding power when subjected to sudden pressure (e.g. staircases) which can cause a thread to pop with each shock. Therefore a twist or spiral shank nail will have the best holding power.

http://www.fourshee.com/history_of_nails.htm

44

THE BIRTH OF PERSONAL COMPUTER

Annika Mombauer

Personal Computers, microcomputers were made possible by two technical innovations in the field of microelectronics: the integrated circuit, or IC, which was developed in 1959; and the microprocessor, which first appeared in 1971. The IC permitted the miniaturization of computer-memory circuits, and the microprocessor reduced the size of a computer's CPU to the size of a single silicon chip.

The invention of the microprocessor, a machine which combines the equivalent of thousands of transistors on a single, tiny silicon chip, was developed by Ted Hoff at Intel Corporation in the Santa Clara Valley south of San Francisco, California, an area that was destined to become known to the world as

Silicon Valley because of the microprocessor and computer industry that grew up there. Because a CPU calculates, performs logical operations, contains operating instructions, and manages data flows, the potential existed for developing a separate system that could function as a complete microcomputer.

The first such desktop-size system specifically designed for personal use appeared in 1974; it was offered by Micro Instrumentation Telemetry Systems (MITS). The owners of the system were then encouraged by the editor of a popular technology magazine to create and sell a mail-order computer kit through the magazine. The computer, which was called Altair, retailed for slightly less than $400.

The demand for the microcomputer kit was immediate, unexpected, and totally overwhelming. Scores of small entrepreneurial companies responded to this demand by producing computers for the new market. The first major electronics firm to manufacture and sell personal computers, Tandy Corporation (Radio Shack), introduced its model in 1977. It quickly dominated the field, because of the combination of two attractive features: a keyboard and a cathode-ray display terminal (CRT). It was also popular because it could be programmed and the user was able to store information by means of cassette tape.

Soon after Tandy's new model was introduced, two engineer-programmers—Stephen Wozniak and Steven Jobs—started a new computer manufacturing company named Apple Computers.

First Apple Computer   In 1976, in what is now the Silicon Valley, Steve Jobs and Steve Wozniak created a homemade microprocessor computer board called Apple I. Working from Jobs' parents' garage, the two men began to manufacture and market the Apple I to local hobbyists and electronics enthusiasts. Early in 1977, Jobs and Wozniak founded Apple Computer, Inc., and in April of that year introduced the Apple II, the world's first personal computer. Based on a board of their design, the Apple II, complete with keyboard and color graphics capability, retailed for $1290.

Some of the new features they introduced into their own microcomputers were expanded memory, inexpensive disk-drive programs and data storage, and color graphics. Apple Computers went on to become the fastest-growing company in U.S. business history. Its rapid growth inspired a large number of similar microcomputer manufacturers to enter the field. Before the end of the decade, the market for personal computers had become clearly defined.

In 1981, IBM introduced its own microcomputer model, the IBM PC. Although it did not make use of the most recent computer technology, the PC was a milestone in this burgeoning field. It proved that the microcomputer industry was more than a current fad, and that the microcomputer was in fact a necessary tool for the business community. The PC's use of a 16-bit microprocessor initiated the development of faster and more powerful micros, and its use of an operating system that was available to all other computer makers led to a de facto standardization of the industry.

In the mid-1980s, a number of other developments were especially important for the growth of microcomputers. One of these was the introduction of a powerful 32-bit computer capable of running advanced multi-user operating systems at high speeds. This has dulled the distinction between microcomputers and minicomputers, placing enough computing power on an office desktop to serve all small businesses and most medium-size businesses.

Another innovation was the introduction of simpler, "user-friendly" methods for controlling the operations of microcomputers. By substituting a graphical user interface (GUI) for the conventional operating system, computers such as the Apple Macintosh allow the user to select icons—graphic symbols of computer functions—from a display screen instead of requiring typed commands. Douglas Engelbart, invented an "X-Y Position Indicator for a Display System": the prototype of the computer "mouse" whose convenience has revolutionized personal computing. New voice-controlled systems are now available, and users may eventually be able to use the words and syntax of spoken language to operate their microcomputers.

45

ASSEMBLY LINE PRODUCTION

Henry Ford and others

A manufacturing technique pioneered in the automobile industry by Henry Ford that lowered production costs and helped bring automobile ownership within the reach of millions of Americans in the early twentieth century. The people behind the invention: Henry Ford (1863-1947), an American carmaker Eli Whitney (1765-1825), an American inventor Elisha King Root (1808-1865), the developer of division of labor Oliver Evans (1755-1819), the inventor of power conveyors

Assembly Line Production

Frederick Winslow Taylor (1856-1915), an efficiency engineer A Practical Man Henry Ford built his first "horseless carriage" by hand in his home workshop in 1896. In 1903, the Ford Motor Company was born. Ford's first product, the Model A, sold for less than one thousand dollars, while other cars at that time were priced at five to ten thousand dollars each. When Ford and his partners tried, in 1905, to sell a more expensive car, sales dropped. Then, in 1907, Ford decided that the Ford Motor Company would build "a motor car for the great multitude." It would be called the Model T. The Model T came out in 1908 and was everything that Henry Ford said it would be. Ford's Model T was a low-priced (about $850), practical car that came in one color only: black. In the twenty years during which the Model T was built, the basic design never changed. Yet the price of the Model T, or "Tin Lizzie," as it was affectionately called, dropped over the years to less than half that of the original Model T. As the price dropped, sales increased, and the Ford Motor Company quickly became the world's largest automobile manufacturer. The last of more than 15 million Model T's was made in 1927. Although it looked and drove almost exactly like the first Model T, these two automobiles were built in an entirely different way. The first was custom-built, while the last came off an assembly line. At first, Ford had built his cars in the same way everyone else did: one at a time. Skilled mechanics would work on a car from start to finish, while helpers and runners brought parts to these highly paid craftsmen as they were needed. After finishing one car, the mechanics and their helpers would begin the next. The Quest for Efficiency Custom-built products are good when there is little demand and buyers are willing to pay the high labor costs. This was not the case with the automobile. Ford realized that in order to make a large number of quality cars at a low price, he had to find a more efficient way to build cars. To do this, he looked to the past and the work of others. He found four ideas: interchangeable parts, continuous flow, division of labor, and elimination of wasted motion. Eli Whitney, the inventor of the cotton gin, was the first person to use interchangeable parts successfully in mass production. In 1798, the United States government asked Whitney to make several thousand muskets in two years. Instead of finding and hiring gunsmiths to make the muskets by hand, Whitney used most of his time and money to design and build special machines that could make large numbers of identical parts—one machine for each part that was needed to build a musket. These tools, and others Whitney made for holding, measuring, and positioning the parts, made it easy for semiskilled, and even unskilled, workers to build a large number of muskets. Production can be made more efficient by carefully arranging the different stages of production to create a "continuous flow." Ford borrowed this idea from at least two places: the meat-packing houses of Chicago and an automatic grain mill run by Oliver Evans. Ford's idea for a moving assembly line came from Chicago's great meat-packing houses in the late 1860's. Here, the bodies of animals were moved along an overhead rail past a number of workers, each of whom made a certain cut, or handled one part of the packing job. This meant that many animals could be butchered and packaged in a single day. Ford looked to Oliver Evans for an automatic conveyor system. In 1783, Evans had designed and operated an automatic grain mill that could be run by only two workers. As one worker poured grain into a funnel-shaped container, called a "hopper," at one end of the mill, a second worker filled sacks with flour at the other end. Everything in between was done automatically, as Evans's conveyors passed the grain through the different steps of the milling process without any help. The idea of "division of labor" is simple: When one complicated job is divided into several easier jobs, some things can be made faster, with fewer mistakes, by workers who need fewer skills than ever before. Elisha King Root had used this principle to make the famous Colt "Six-Shooter." In 1849, Root went to work for Samuel Colt at his Connecticut factory and proved to be a manufacturing genius. By dividing the work into very simple steps, with each step performed by one worker, Root was able to make many more guns in much less time. Before Ford applied Root's idea to the making of engines, it took one worker one day to make one engine. By breaking down the complicated job of making an automobile engine into eighty-four simpler jobs, Ford was able to make the process much more efficient. By assigning one person to each job, Ford's company was able to make 352 engines per day—an increase of more than 400 percent. Frederick Winslow Taylor has been called the "original efficiency expert." His idea was that inefficiency was caused by wasted time and wasted motion. So Taylor studied ways to eliminate wasted motion. He proved that, in the long run, doing a job too quickly was as bad as doing it too slowly. "Correct speed is the speed at which men can work hour after hour, day after day, year in and year out, and remain continuously in good health," he said. Taylor also studied ways to streamline workers' movements. In this way, he was able to keep wasted motion to a minimum. Impact The changeover from custom production to mass production was an evolution rather than a revolution. Henry Ford applied the four basic ideas of mass production slowly and with care, testing each new idea before it was used. In 1913, the first moving assembly line for automobiles was being used to make Model T's. Ford was able to make his Tin Lizzies faster than ever, and his competitors soon followed his lead. He had succeeded in making it possible for millions of people to buy automobiles. Ford's work gave a new push to the Industrial Revolution. It showed Americans that mass production could be used to improve quality, cut the cost of making an automobile, and improve profits. In fact, the Model T was so profitable that in 1914 Ford was able to double the minimum daily wage of his workers, so that they too could afford to buy Tin Lizzies. Although Americans account for only about 6 percent of the world's population, they now own about 50 percent of its wealth. There are more than twice as many radios in the United States as there are people. The roads are crowded with more than 180 million automobiles. Homes are filled with the sounds and sights emitting from more than 150 million television sets. Never have the people of one nation owned so much. Where did all the products—radios, cars, television sets—come from? The answer is industry, which still depends on the methods developed by Henry Ford.

46

THE COTTON GIN

Eli Whitney

The third best known American inventor of the pre-atomic age, after Thomas Edison and Alexander Graham Bell, is probably Eli Whitney. Whitney certainly transformed the economies of the antebellum North and South. But among invention aficionados, his invention of the cotton gin is a matter of some dispute.

Whitney was born in Westboro, Massachusetts in 1765. As a child, he showed an instinct and talent for machinery. He worked as a blacksmith, and invented a nail-making machine. Whitney's dream of attending Yale College was frustrated for some years, because no college then taught or much appreciated the "useful arts." But Whitney did attend Yale, and graduated at the age of 27, only to find that there were no jobs for engineers either. So he accepted a teaching position in South Carolina.

En route, in early 1793, Whitney was befriended by Katherine Greene, the widow of a Revolutionary War general. When Whitney's teaching job later fell through, Greene invited him to stay at her plantation, Mulberry Grove, where she thought he might make himself helpful. As Whitney soon discovered, most cotton plantations were then on the brink of insolvency, because "green seed" cotton, the only strain that would grow inland, took too long to cull from its seeds. To sift out a single "point" of cotton lint from its surrounding seeds required ten hard hours of hand labor.

Everyone agreed that the solution was a machine to do this work; but no one had been able to make one. According to legend, within ten days of his arrival Whitney had observed the manual process and built a machine that did the same thing much faster. It is clear that his very first model did not work. In it, the bulk cotton was pressed against a wire screen, which held back the seeds while wooden teeth jutting out from an adjacent rotating drum teased the cotton fibers out through the mesh. This model invariably jammed. The next version was a complete success, thanks to thin wire hooks replacing the wooden teeth, and a moving brush that constantly cleared away the collected fibers.

By all accounts, Greene encouraged Whitney. The vexed question is whether the key element, the wire hooks, was his idea or hers. Greene supporters cite the claim of a friend of a friend of her plantation foreman, that Greene invoked "a woman's wit" and told Whitney to replace his wooden pegs with the wires of a fireplace cleaning brush. Whitney supporters cite a letter to the editor of Southern Agriculturalist magazine, whose author heard from admittedly shadowy sources that Whitney had explicitly asked Greene for a pin to experiment with at the start of his efforts. (Note that for some time during his Massachusetts days, Whitney had been the New World's sole manufacturer of hatpins.)

Whatever the comparative contributions, the cotton gin ("gin" is simply short for "engine") was a stupendous success. After Whitney gave a one-hour demonstration, in which the machine did the day's work of many men, farmers raced to sow their fields with green seed cotton. As the cotton grew, Whitney's workshop was broken into and his machine was examined in detail: soon, copies were everywhere. Whitney could not possibly have manufactured one tenth of the gins that that first crop would require; but it is nonetheless unfair that his patent (granted in 1794) guaranteed him only ten years of legal battles, which ended in penury.

In 1804, Whitney left the South forever, disappointed and disgusted. In his words, "An invention can be so valuable as to be worthless to the inventor." In fact, Whitney never attempted to patent any of his later inventions (for example, a milling machine). But after settling in New Haven, Connecticut, Whitney re-invented American manufacturing as a whole, through mass production.

Whitney wanted to enable unskilled laborers to make complex products. He managed this by designing products (his test case was rifles) with interchangeable parts. These were cut and shaped by machines that each performed one precise function over and over again. The workers would merely put each machine through its motions.

Mass production is not a romantic notion. But it allowed for an unprecedented boom in American industry, and eventually provided employment for thousands of workers who were unwilling or unable to acquire apprenticeships in skilled crafts. And by all accounts, Eli Whitney himself treated his "manufactory" workers with appreciation and respect: the awful abuses of laborers that came about after his death in 1825 were a perversion of his system.

http://web.mit.edu/invent/iow/whitney.html

47

Hygiene and Sanitation

The word hygiene comes from Hygeia, the Greek goddess of health (photo, below), who was the daughter of Aesculapius, the god of medicine. Since the advent of the Industrial Revolution (c.1750-1850) and the discovery of the germ theory of disease in the second half of the nineteenth century, hygiene and sanitation have been at the forefront of the struggle against illness and disease.

Together with the great strides made in improvements in the standards of living provided by free market capitalism, economic freedom, and the advances in scientific medicine --- hygiene and sanitation have resulted in unprecedented longevity, concomitant with markedly improved quality of life in the last century and a half of medical history.

Thanks to these advances, senior citizens, particularly octogenarians, have become the fastest growing segment of our population even though the priority assigned to the prolongation of life span has taken a back seat to other items in health care policy, chiefly the containment of health care costs and "the proper allocation of finite and scarce health resources." Thus, the concept of longevity (from the Latin longaevitas, meaning "long-lived") has been almost abandoned for the new, modern concerns of "useful life span," "the duty to die," "assisted suicide," and so on.

Nevertheless the dramatic extension of life span closely associated with improvement in the quality of life is welcomed news for the American "baby boomers," who have the most to gain from advances in longevity as they age in the first half of the twenty-first century.

In the Middle Ages, the average human life expectancy did not reach into the teen years, not only because of the extremely high perinatal mortality that heavily skewed the data, but

Hygiene and Sanitation

also because Europeans (and much of the world during this time) lived in an unhealthy milieu of filth, poor hygiene, and nearly non-existent sanitation. Superstition and ignorance, along with pestilential diseases and vermin infestation, were rampant. Epidemic and endemic diseases such as the bubonic plague, typhus, variola (smallpox), and the White Death of tuberculosis (consumption) took a heavy toll on the population, both young and old.

Sanitation in the Middle Ages, from an old wood cut

During the Middle Ages until the mid-nineteenth century cleanliness was just not a priority. The streets in those days were dumping grounds for refuse, and domestic animals including hogs roamed the streets. According to medical historian Howard W. Haggard: "Refuse from the table was thrown on the floor to be eaten by the dog and cat or to rot among the rushes and draw swarms of flies from the stable. The smell of the open cesspool in the rear of the house would have spoiled your appetite, even if the sight of the dining room had not."(2) There was little improvement in this dire, unhealthy milieu until the mid- to late nineteenth century when the advances of the aforementioned Industrial Revolution and the discovery of the germ theory of disease brought about public health measures that, building upon the importance of good hygiene and sanitation, culminated in the rise of the scientific era of medicine. The heroes and heroines of this age included such notable medical figures as: Edward Jenner (1749-1823), Oliver Wendell Holmes (1809-1894), Ignaz Semmelweiss (1818-1865; photo, right below), Florence Nightingale (1820-1910; photo, left below), Rudolf Virchow (1821-1902), Clara Barton (1821-1912), Louis Pasteur (1822-1895), Joseph Lister (1827-1912), J. Henri Dunant (1828-1910), and Robert Koch (1843-1910).

In the words of the surgeons and medical writers Nathan Hiatt and Jonathan R. Hiatt, "The industrial revolution, however, also brought a raised standard of living, with higher wages, improved nutrition, cheap soap, and inexpensive cotton clothing. Cotton clothing, unlike the louse-ridden woolens worn in the past, could be and had to be washed, thus dispossessing lice and helping to end typhus epidemics. By 1900, improved nutrition, better sanitation, and, especially, contributions from bacteriologists increased life expectancy at birth by almost six years (to age 47.3)..."

Of particular importance in medical history, puerperal fever was one of those diseases that intrigued and baffled doctors in the nineteenth century. You might even remember the famous painting of the illustrious Dr. Oliver Wendell Holmes delivering his famed lecture on the subject to the Boston Medical Society in 1843. Just as Dr. Semmelweiss had predicted, the disease was conquered when obstetricians began washing their hands between deliveries. Puerperal fever was eradicated with cleanliness. Likewise, surgical mortality became acceptable when surgeons began washing their hands and using antiseptic techniques as urged by Dr. Joseph Lister. The scientific tenets of bacteriology and microbiology introduced by Louis Pasteur were finally being applied to obstetrics, medicine and surgery.

The engine behind the drive for hospital reform in the mid-nineteenth century was Florence Nightingale (photo, left). After her tremendously successful humanitarian venture at the Scutari Barrack Hospital during the Crimean War, Nightingale was able to convince the world of the necessity of improving hygiene and sanitation as well as having trained professional nurses tending the sick in the hospital wards. According to medical historian Guy Williams, when she arrived at Scutari "there were plenty of rats, lice and fleas, but there were very few knives, forks, or spoons. Miss Nightingale and her nurses, who were allowed just one pint of water per person per day for washing and drinking and for making tea, [yet]...the ladies' own personal circumstances were hardly hygienic."(4) With hard work and determination, she turned the situation around and by the time she returned to England, she had become a national heroine.

Maternal mortality, a dreaded and common complication of pregnancy throughout the ages, was all but conquered in the West in the twentieth century by a three-pronged attack of public health, particularly the efforts at better hygiene and sanitation; improved obstetrical care, and the use of antibiotics.

The period between 1930 and 1940 saw a sharply rising curve in longevity rates thanks to the widespread usage of antibiotics and the much improved standards in cleanliness, hygiene, and sanitation. Thereafter, further reductions in maternal and infant mortalities were to a significant degree responsible for the tremendous rise in life expectancy. With the conquest of such diseases and scourges of humanity as syphilis, pneumonia, diphtheria, typhoid fever, typhus , and earlier in the century, the old consumptive killer, tuberculosis --- life expectancy climbed from 59.7 years in 1930 to 74.9 years by 1987.

By the 1980s, the widespread availability and use of sulfa drugs and penicillin atop earlier traditional public health measures prolonged life beyond all expectations. These traditional health measures included: isolation of the sick during epidemics; quarantining of ships at ports of disembarkation; disinfection of fomites; exposure to fresh air and the beneficial rays of sunlight; and widespread immunization practices. The impact of these measures was enhanced by education and promotion of personal hygiene and communal sanitation, including the use of potable, running water and the proper disposal of wastes.
With the avoidance of self-destructive behavior, cessation of smoking, maintenance of ideal body weight, proper regime of exercises, adequate control of blood pressure and cholesterol levels, proper management of stress, and so on, one can still stretch his or her life span considerably in the twenty-first century.

Consider that over 80 percent of diseases are associated with unhealthy lifestyles and self-destructive behaviors and thus are subject to healthy alterations in behavior.(5) Needless to say, as the author has pointed out elsewhere, the possibilities for improvement become enormous. Maximal life span, redolent of the search for the fountain of youth by Ponce de León, has been estimated to be 114 years. Thus, there is still room for improvement.

Protecting our health can even reduce health care costs and save money in the process. The money saved can then be spent when we reach a ripe old, antediluvian age, when most of us have reached our personal best in terms of knowledge and wisdom!

http://www.haciendapublishing.com/medicalsentinel/medical-history-hygiene-and-sanitation

A Short History of Steel: Part II

Terence Bell

The Bessemer Process and Modern Steelmaking

The growth of railroads during the 19th century in both Europe and America put great pressure on the iron industry, which still struggled with inefficient production processes. Yet steel was still unproven as a structural metal and production

was slow and costly. That was until 1856, when Henry Bessemer came up with a more effective way to introduce oxygen into molten iron in order to reduce the carbon content.
Now known as the Bessemer Process, Bessemer designed a pear-shaped receptacle - refered to as a 'converter' - in which iron could be heated while oxygen could be blown through the molten metal. As oxygen passed through the molten metal, it would react with the carbon, realasing carbon dioxide and producing a more pure iron.

The process was fast and inexpensive, removing carbon and silicon from iron in a matter of minutes but suffered from being too successful. Too much carbon was removed and too much oxygen remained in the final product. Bessemer ultimately had to repay his investors until he could find a method to increase the carbon content and remove the unwanted oxygen.

At about the same time, British metallurgist Robert Mushet acquired and began testing a compound of iron, carbon and manganese - known asspeigeleisen. Manganese was known to remove oxygen from molten iron and the carbon content in the speigeleisen, if added in the right quantities, would provide the solution to Bessemer's problems. Bessemer began adding it to his conversion process with great success.
Yet, one problem still remained. Bessemer had failed to find a way to remove phosphorus - a deleterious impurity that makes steel brittle - from his end product. Consequently, only phosphorus-free ores from Sweden and Wales could be used.
In 1876 Welshman Sidney Gilchrist Thomas came up with the solution by adding a chemically basic flux - limestone - to the

Bessemer process. The limestone drew phosphorus from the pig iron into the slag, allowing the unwanted element to be removed.

This innovation meant that, finally, iron ore from anywhere in the world could be used to make steel. Not surprisingly, steel production costs began decreasing significantly. Prices for steel rail dropped more than 80% between 1867 and 1884, as a result of the new steel producing techniques, initiating growth of the world steel industry.

The Open Hearth Process:

In the 1860s German engineer Karl Wilhelm Siemens further enhanced steel production through his creation of the open hearth process. The open hearth process produced steel from pig iron in large shallow furnaces.

Using high temperatures to burn off excess carbon and other impuriites, the process relied on heated brick chambers below the hearth. Regenerative furnaces later used exhaust gases from the furnace to maintain high temperatures in the brick chambers below.

This method allowed for the production of much larger quantities (50-100 metric tons could be produced in one furnace), periodic testing of the molten steel so that it could be made to meet particular specifications and the use of scrap steel as a raw material. Although the process itself was much slower, by 1900 the open hearth process had largely replaced the Bessemer process.

A Short History of Steel

Birth of the Steel Industry:
The revolution in steel production that provided cheaper, higher quality material, was recognized by many businessmen of the day as an investment opportunity. Capitalists of the late 19th century, including Andrew Carnegie and Charles Schwab, invested and made millions (billions in the case of Carnegie) in the steel industry. Carnegie's US Steel Corporation, founded in 1901, was the first corporation ever launched valued at over one billion dollars.

Electric Arc Furnace Steelmaking
Just after the turn of the century, another development occurred that would have a strong influence on the evolution of steel production. Paul Heroult's electric aric furnace (EAF) was designed to pass an electric current through charged material, resulting in exothermic oxidation and temperatures up to 3272°F (1800°C), more than sufficient to heat steel production.

Initially used for specialty steels, EAFs grew in use and, by World War II, were being used for the manufacturing of steel alloys. The low investment cost involved in setting up EAF mills allowed them to compete with the major US producers like US Steel Corp. and Bethlehem Steel, especially in carbon steels, or long products.

Because EAFs can produce steel from 100% scrap - or cold ferrous - feed, less energy per unit of production is needed. As opposed to basic oxygen hearths, operations can also be stopped and started with little associated cost. For these reasons, production via EAFs has been steadily increasing for over 50 years and now accounts for about 33% of global steel production.

Oxygen Steelmaking

The majority of global steel production - about 66% - is now produced in basic oxygen facilities. The development of a method to separate oxygen from nitrogen on an industrial scale in the 1960s allowed for major advances in the development of basic oxygen furnaces.
Basic oxygen furnaces blow oxygen into large quantities of molten iron and scrap steel and can complete a charge much more quickly than open hearth methods. Large vessels holding up to 350 metric tons of iron can complete conversion to steel in less than one hour.

The cost efficiencies of oxygen steelmaking made open hearth factories uncompetitive and, following the advent of oxygen steelmaking in the 1960s, open-hearth operations began closing. The last open-hearth facility in the US closed in 1992 and in China in 2001.

http://metals.about.com/od/properties/a/A-Short-History-Of-Steel-Part-Ii.htm

49

THE STEAM TURBINE

Sir Chales Parsons

Since the amenities of civilised life depend almost entirely on the availability of power for industrial purposes, those pioneers who have provided mankind with the means of obtaining power more cheaply and abundantly will always rank high among the benefactors of humanity. From this point of view no man has made a greater contribution to human welfare than Sir Charles Parsons by the revolutionary improvements he brought about in the use of steam. His name will always be particularly associated in the minds of the public with the invention of the steam turbine and its application to duties on land and sea, although, as will be seen, his contribution to the advance of science and engineering extended far beyond the limits of a single invention. Even if Parsons had done nothing more than produce the first practical steam turbine his fame as an engineer would have been secure for all time. By its introduction he exercised an influence upon industry that was comparable only with that of Watt about a century earlier, though vastly more far-reaching in its effects by reason of the wider field open to him.

When Watt built his first condensing steam engines, the operation of pumping machinery for mines was almost the only duty for which engines were required. Towards the end of the eighteenth century steam engines began to take the place of water wheels as prime movers for mills and factories, but as the dynamo had not then been invented, the generation of electricity and all the industrial development that depends on it lay still in the future. Nor was there at the time, except perhaps in the minds of a few enthusiasts, any idea that steam would ever be used for the propulsion of ships. There is no evidence that Watt foresaw the immense field there would be for steam power in marine work, even though his firm of Boulton and Watt constructed the engine that drove Fulton's historic little steam vessel the Clermont in 1806. Parsons, on the other hand, commenced his life's work when the two great branches of electrical and marine engineering were already established, each of them offering an unbounded scope for the steam turbine as soon as its practicability could be demonstrated. In another respect also the times were auspicious for him. The reciprocating steam engine, which had held the field unchallenged for a hundred years, had practically reached the limit of its development. The labours of generations of engineers had raised it to a very high degree of excellence, and no further refinement of design was capable of effecting any substantial increase in its efficiency. Furthermore, the powers for which reciprocating engines could be built were restricted by considerations of size and weight to units of a few thousand horse-power only, so that if any further advance was to be made in steam engineering it could only take place along some totally different lines. It is as true in engineering as in every other kind of evolutionary activity, that approach to perfection in any direction is equally the approach to stagnation, so that if progress is to continue some radical departure has to be made.

THE STEAM TURBINE

Parsons, alone among his contemporaries, saw in the turbine principle the means of escape from the limitations of the reciprocating engine, or perhaps it would be more accurate to say that he alone possessed the genius and courage to transform a possibility into a reality. That he foresaw from the outset the wide diversity of the duties to which the turbine could be applied is clear from his earliest patents, which also showed a remarkable understanding of the conditions essential to success in meeting the requirements of each particular case. Problem after problem was solved in the most admirable way, and, instead of being merely an ingenious toy, as many people at first considered it, the turbine steadily and surely won recognition as the standard type of prime mover wherever the production of steam power was concerned. It can be constructed for far larger outputs than any reciprocating engine, and it is moreover much more economical of steam. Indeed, apart from the benefits that the work of Parsons has conferred upon the present generation, the economy which he made possible in the consumption of the exhaustible fuel resources of the world entitles him to the gratitude of posterity.

For those who are not acquainted with the principle of the steam turbine, it may be well to explain briefly the nature of the great invention of Parsons. The object that he set himself was that of producing power by utilising the velocity of a jet of steam, instead of using the pressure of the steam to drive a piston as in the ordinary reciprocating engine. It was evident that a jet of steam could be made to turn a wheel by acting on blades set around its circumference, or alternatively it could be used to develop power by its own reaction when escaping tangentially from an orifice in a rotating wheel or arm. Both devices had already been suggested by innumerable inventors, but the hitherto insuperable difficulty in constructing a practical turbine by either method lay in utilising the excessive velocity of the steam. Even steam at a comparatively low pressure escaping into the atmosphere may easily be travelling at more than 2500 feet per second, or over 1700 miles an hour, while twice this velocity may be attained by high-pressure steam flowing into a good vacuum. To make use of such velocities effectively in a simple turbine, the blades or other moving elements would have to travel at about half the speed of the steam, for otherwise an undue proportion of the energy of the jet would be uselessly carried away in the steam leaving the wheel. The blade speeds required for efficiency would therefore be so high that they would be prohibited by reason of centrifugal force alone, apart from other considerations. That Watt, with his sound engineering instinct, had appreciated this fact' is shown by one of his letters to Boulton. His partner had expressed fears as to the effect that the competition of a proposed steam turbine might have on their engine-building business, but Watt had disposed of them with the remark that 'Without God makes it possible for things to move 1000 feet per second, it cannot do much harm'.

Although there are to-day large turbines containing blades whose tips travel at speeds even greater than Watt thought would be possible only by a special dispensation of Providence, such speeds were out of the question under the conditions existing when Parsons commenced his work. He could therefore only secure a proper relationship between steam speed and blade speed by reducing the former to a manageable amount. Now the speed of a jet of steam will obviously depend upon the difference of pressure that causes the flow. It occurred to Parsons that he could attain his end by the device of causing the whole expansion of the steam to take place by a series of steps, each partial drop of pressure being only sufficient to generate a velocity that could be efficiently utilised by blades running at a moderate speed. To put this idea into effect he constructed a turbine consisting of a cylindrical rotor enclosed in a casing. The steam flowed along the annulus between the two, parallel to the axis of the machine, and in so doing it had to pass through rings of blades fixed alternately in the casing and rotor. The passages between the blades of each ring formed virtually a set of nozzles in which a partial expansion of the steam could take place. In passing through each ring of fixed blades the steam acquired a certain velocity due to this expansion, and the jets so formed gave up their energy in driving the succeeding row of moving blades. The passages between the latter blades also acted as nozzles, permitting a further partial expansion, so that the moving blades were impelled partly by the 'action' of the steam entering them and partly by the 'reaction' of the steam leaving them.

The principle of subdividing the whole expansion of the steam into a number of stages, so that only comparatively moderate velocities have to be dealt with, still forms the basis of all efficient turbine design. The secondary principle of utilising the 'reaction' of the steam expanding in moving blades has remained typical of the Parsons turbine. It is not, however, an indispensable characteristic of an efficient turbine, and certain inventors subsequent to Parsons, notably C. G. Curtis in the United States and Professor A. Rateau in France, preferred for constructional reasons to confine the expansion of the steam to fixed nozzles. Machines of the latter type, in which the steam drives the blading of each stage by virtue of its velocity only, are known as 'impulse' turbines. Although they have attained an honourable position in the industry, it is generally recognized that the 'reaction' principle, chosen by Parsons for his original turbine, is conducive to the highest efficiency, so that large machines which are nominally of the impulse type are now often designed to work with a certain amount of reaction in their blading.

In addition to laying down the broad lines necessary to success in the development of the new kind of prime mover, Parsons had many practical problems to solve before his ideas could be embodied in an actual machine. Not only had a suitable form of blading to be invented and appropriate manufacturing methods devised, but the design generally had to conform to conditions quite outside the range of ordinary engineering practice. For example, to obtain the desired blade velocity in the small turbine he first constructed a rotational speed of no ,000 revolutions per minute had to be adopted. This was over fifty times as fast as the fastest reciprocating engine of the day, and It involved the invention of a new kind of bearing which would permit of a long rotor, inevitably out of mathematically perfect balance, running at such a speed without vibration. Means had also to be provided for the continuous lubrication of these bearings, and a totally new method of controlling the speed of the machine had to be devised. Again, it was realised that the flow of the steam would result in an end-thrust on the blading, and to prevent this being transmitted to the bearings, where it might have caused trouble, Parsons neutralised it by the ingenious expedient of admitting the steam midway along the rotor and causing it to flow equally towards each end. His subsequent invention of 'dummy pistons' rendered the double flow principle unnecessary for machines of moderate output, but without it the large and efficient high-speed machines of to-day could hardly be built. A study of Parsons' first turbine patent, taken out in 1884, will show how clearly he appreciated the difficulties in his path and how thoroughly he had considered the means of overcoming them. Again, obvious as the principle of expanding the steam by stages now appears to us, the invention must be regarded in the light of the state of the art at the time. The only previous attempts to develop any useful amount of power from steam, otherwise than by causing it to drive a piston, had taken the form of machines driven by the reaction of jets issuing from the ends of rotating arms, on the lines of the classical Aeropile of Hero of Alexandria. The famous Cornish engineer, Richard Trevithick, had constructed what he called a "whirling engine' on this principle in 1815, and other more or less workable machines of the same kind had been made from time to time, but the lessons to be drawn from them were rather of warning than encouragement. It is true that various inventors had propounded plans for the more rational utilisation of steam in machines of the turbine type, but no such machines had assumed a practical form and steam engineers in general believed that any attempt to supersede the reciprocating engine as a prime mover was foredoomed to failure. The success of Parsons' first little turbine marked the beginning of the most revolutionary change in the history of steam engineering. By developing power from the velocity of steam rather than from its static pressure, the turbine was exempt from the mechanical limitations of the reciprocating engine. Its invention has enabled the power that could be produced by a given weight and size of machinery to be multiplied a hundredfold and it has provided that purely rotational motion at high speed so desirable for the driving of electrical generators and many other classes of machinery. In addition to these advantages, it has brought about a remarkable economy in the use of steam.

http://www.houseofdavid.ca/parsons.htm

50

THE HUMAN GENOME: A NEW REALITY

In June 1985, as dusk encroached on the second millennium, meetings aimed at outlining the practical task of sequencing the human genome began at the University of California, Santa Cruz. The scientific and technological conditions of the 1980s had become a catalyst for these discussions. DNA cloning and Fred Sanger's sequencing methods, developed in the mid- to late 1970s, were being exploited by scientists who felt that sequencing the human genome seemed possible at an experimental level. Crucially, researchers were, at the same time, beginning to apply computing solutions to genetics and DNA sequencing, developing methods that would make feasible the task of generating and handling genetic data globally.

The Human Genome: A New Reality

This grand, new concept - a "Human Genome Project" - had strong supporters, who argued that deciphering the human genome would lead to new understanding and benefits for human health as well as and determined detractors, who feared such a project would provide a product that would bear little explanatory power for humans - perhaps merely a meaningless string of letters. Even before the Human Genome Project began in earnest, some commentators feared that this project had "engendered a controversy... that involves personalities and politics."
The personalities, the politics and the controversy were only just emerging.

[Morag Lewis, Genome Research Limited]

The Human Genome Project launched in 1990, through funding from the US National Institutes of Health (NIH) and Department of Energy, whose labs joined with international collaborators and resolved to sequence 95% of the DNA in human cells in just 15 years. Meanwhile in the UK, John Sulston and his colleagues at the MRC's Laboratory of Molecular Biology in Cambridge, had, for several years, been working at mapping the genome of the nematode worm and had resolved that sequencing the entire genome of the worm was finally feasible. As the Human Genome Project was progressing in the US, in the UK the MRC approached the Wellcome Trust suggesting they form a new partnership to fund John's proposed worm sequencing, as a pilot for the Human Genome Project. From here things soon snowballed: the Wellcome Trust suggested that a much larger sequencing effort, to bolster the Human Genome Project should be embarked upon in the UK and appointed one of their senior administrators, Michael Morgan, to look into the viability of such a sequencing initiative. Eventually, in 1992, John Sulston submitted a grant application for an enormous £40-50 million to fund a new centre - the Sanger Centre - which was to form the British arm of the Human Genome Project's sequencing efforts.

In 1993 - with funding from the Wellcome Trust and MRC - the Sanger Centre was officially opened. One scientist recalls being struck by the scale of the task that lay ahead, on arriving at the Institute in 1993 Simon Gregory reflects: "it was just a huge lab, a huge empty lab, with boxes and boxes of equipment. It was all very exciting."

By the end of that year 87 scientists were working at the Sanger Centre, under the leadership of John Sulston, beginning to map and sequence the human genome.

The global sequencing effort

To sequence the human genome as accurately as possible, researchers developed the 'hierarchical shotgun' method. Researchers agreed that this was the best way to achieve the Human Genome Project's target of 95% coverage of the human genome by 2005.

[Morag Lewis, Genome Research Limited]

The first challenge was to create a map of the human genome - a set of index marks on the genome code, used to position the sequences of letters of code that would come later.

Researchers essentially broke many copies of the genome into fragments, each around 150,000 letters of code (or base-pairs) long. They inserted the fragments into a bacterial artificial chromosome that could be grown in E. coli bacteria which divided, thereby replicating the DNA samples to create a stable resource - a 'library' of DNA clones. Where the cloned fragments came from or which overlapped was not known at this point.

The Human Genome: A New Reality

[Morag Lewis, Genome Research Limited]

Using special enzymes, researchers could cut the individual clones into diagnostic 'fingerprint' of fragments defined by each clone's sequence. They could then search among millions of fingerprints for shared fragments that would reveal overlaps among the clones. Researchers then assembled the clones into longer contiguous regions and mapped these onto the human chromosomes. The result: a physical human genome map that would be crucial for the sequencing efforts.

To generate sequence of the individual bases that make up the genome, scientists needed to break the cloned fragments into smaller, more manageable, chunks, each around 1000 to 2000 base-pairs long. Researchers sequenced these fragments of human DNA using the shotgun method developed by Fred Sanger and his colleagues a dozen years before. Much as in mapping, researchers used overlaps, this time in the letters of

genetic code itself, to reassemble the short stretches of determined sequence. Assembling the sequence from many short segments of sequence was a hugely intense compute task that depended on emerging technology and software to succeed.
Gradually labs around the world began producing DNA sequence. By 1994, the Sanger Institute had produced its first 100,000 bases of human DNA sequence. Remarkably, researchers at the Institute had already produced ten times that amount from the nematode worm genome. The worm project was a trailblazer- its methods, practices, collaborations and ethos would be integral to the development the social mores that would later lead to the successful completion of the Human Genome Project.

As the human sequence data was pouring out from centres across the globe, researchers were afforded glimpses of the kind of power that the human genome sequence might have for medical advance. In 1995, researchers from the Sanger Centre, with international collaborators, located theBRCA2 gene, associated with increased risk of breast cancer. Elsewhere, as early as 1993, a US team had located the MSH2 gene, which increases the risk of colon cancer for carriers. In Canada, researchers found five variants on the FAD gene, which together confer an almost 100 per cent risk of developing Alzheimer's disease.

https://www.sanger.ac.uk/about/history/hgp/